185
Advances in Biochemical Engineering/Biotechnology

Aims and Scope

This book series reviews current trends in modern biotechnology and biochemical engineering. Its aim is to cover all aspects of these interdisciplinary disciplines, where knowledge, methods and expertise are required from chemistry, biochemistry, microbiology, molecular biology, chemical engineering and computer science.

Volumes are organized topically and provide a comprehensive discussion of developments in the field over the past 3–5 years. The series also discusses new discoveries and applications. Special volumes are dedicated to selected topics which focus on new biotechnological products and new processes for their synthesis and purification.

In general, volumes are edited by well-known guest editors. The series editor and publisher will, however, always be pleased to receive suggestions and supplementary information. Manuscripts are accepted in English.

In references, Advances in Biochemical Engineering/Biotechnology is abbreviated as *Adv. Biochem. Engin./Biotechnol.* and cited as a journal.

Yuan Lu • Michael C. Jewett
Editors

Cell-free Macromolecular Synthesis

With contributions by

M. Cao · A. L. Cortajarena · J. Gu · T. Huber · L. Kai · J. Koo · J. Li · J. Li · P. Li · Y. Li · F. López-Gallego · A. Manteca · H. Qi · M. Sánchez-Costa · L. Shen · R. Sieskind · D. Van Raad · Y. Yang · K. Yue

Springer

Editors
Yuan Lu
Tsinghua University
Beijing, China

Michael C. Jewett
Stanford University
Stanford, CA, USA

ISSN 0724-6145 ISSN 1616-8542 (electronic)
Advances in Biochemical Engineering/Biotechnology
ISBN 978-3-031-41286-8 ISBN 978-3-031-41287-5 (eBook)
https://doi.org/10.1007/978-3-031-41287-5

This Springer imprint is published by the registered company Springer Nature Switzerland AG
The registered company address is: Gewerbestrasse 11, 6330 Cham, Switzerland

Paper in this product is recyclable.

This book is dedicated to James "Jim" R. Swartz. Jim is the James H. Clark Professor in the School of Engineering and Professor of Chemical Engineering and of Bioengineering at Stanford University. He is a world leader of the large, vibrant, and international biochemical engineering community and a founder of the cell-free biotechnology field. Jim's fundamental, original, and pioneering contributions to the field, including seminal bioprocess development innovations, enabled the dawn of a new era in cell-free protein synthesis. This led to a new sector in the biotech industry that is transforming production of protein therapeutics, personalized medicines, vaccines, diagnostics, and value-added biochemicals.

Preface

Cell-free macromolecular synthesis is the technical core of cell-free synthetic biology, also known as in vitro protein translation, which is a technique used to supplement cell-based protein expression. Cell-free systems are simple to operate and easy to control, and their advantages over protein expression in vivo include their open nature, elimination of dependence on living cells, and concentration of all system material and energy on target protein production. Therefore, cell-free protein synthesis, as a platform technology, is expected to overcome the expression limitations caused by cell membrane constraints in the current intracellular production system and has broad prospects in basic and applied science research.

Cell-free protein synthesis has significant and outstanding advantages in the following aspects. The system is open, can tolerate cytotoxic proteins, and can produce proteins that are difficult for cells to express, such as membrane proteins and complex proteins. The system's reaction is rapid, and the eukaryotic cell-free system has abundant post-translational modification of proteins. It is easy to regulate the reaction process and facilitate the site-specific binding of non-natural amino acids to proteins to synthesize non-natural proteins. Biosensors based on cell-free systems are easy to operate, can solve the problem of biosafety, and produce a wide variety of sensors. Cell-free systems on paper that can be combined with freeze-drying technology can realize portable reactions, break the cold chain transport problem, and store cell-free reagents to quickly produce high-titer vaccines. Because of these advantages, cell-free systems have great potential for expressing recombinant functional proteins, modified proteins, non-natural proteins, portable biosensors, and in vitro diagnostics.

Because of the openness, controllability, efficiency, flexibility, and many other advantages, cell-free protein synthesis system as a powerful synthetic biology technology platform has been diversified and vigorously developed. Its application is across biological manufacturing, biocatalysis, biomedicine, biosensing, artificial cells, and other important basic and applied research fields. Various cell-free biosynthesis technology platforms further enable in vitro production of proteins of varying complexity and species origin. However, the development of cell-free protein synthesis systems still faces many challenges. In future development, cell-free systems need to be further optimized to increase efficiency, reduce costs, and improve the personalization, diversification, universality, and stability of the synthesis of biological macromolecules (RNA and protein). The half-life of cell-free systems needs to be further extended, and it is necessary to move toward cell-free synthesis systems that can realize self-replication. Multidisciplinary fields such as advanced materials science, artificial intelligence, and biomedicine must be further integrated to show the potential for broader applications of cell-free synthetic systems.

Welcome to this comprehensive collection of chapters about the intriguing field of cell-free macromolecular synthesis. Whether you are an experienced expert in cell-free synthetic biology or a newcomer to academia, these resources will be extremely helpful. These chapters cover important current aspects, including advanced tools, novel protein design, molecular evolution, and cutting-edge techniques. We are confident that this information will inspire and empower you as you pursue greater knowledge and advancements.

Lastly, we want to express our gratitude to all the authors and editors for their valuable contributions. Without their hard work and dedication, this project would not have been possible. We appreciate their input and look forward to working with them again in the future.

Beijing, China Yuan Lu
Stanford, CA, USA Michael C. Jewett

Contents

Adv Biochem Eng Biotechnol (2023) 185: 1–20
https://doi.org/10.1007/10_2023_232

Published online: 2 August 2023

Bottom-Up Synthetic Biology Using Cell-Free Protein Synthesis

Ke Yue, Yingqiu Li, Mengjiao Cao, Lulu Shen, Jingsheng Gu, and Lei Kai

Contents

Abstract Technical advances in biotechnology have greatly accelerated the development of bottom-up synthetic biology. Unlike top-down approaches, bottom-up synthetic biology focuses on the construction of a minimal cell from scratch and the application of these principles to solve challenges. Cell-free protein synthesis (CFPS) systems provide minimal machinery for transcription and translation, from either a fractionated cell lysate or individual purified protein elements, thus speeding up the development of synthetic cell projects. In this review, we trace the history of the cell-free technique back to the first in vitro fermentation experiment using yeast cell lysate. Furthermore, we summarized progresses of individual cell mimicry modules, such as compartmentalization, gene expression regulation, energy regeneration and metabolism, growth and division, communication, and motility. Finally, current challenges and future perspectives on the field are outlined.

K. Yue, Y. Li, M. Cao, L. Shen, J. Gu, and L. Kai (✉)
School of Life Sciences, Jiangsu Normal University, Xuzhou, China
e-mail: lkai@jsnu.edu.cn

Graphical Abstract

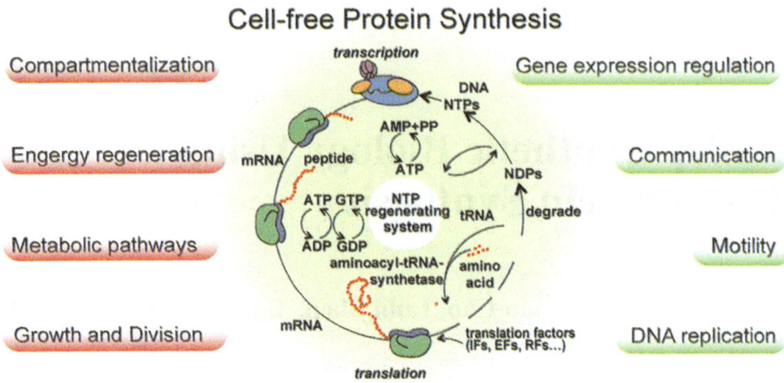

Keywords Bottom-up synthetic biology, Cell-free protein synthesis, Minimal cells, Synthetic cells

1 Introduction

The scope of synthetic biology today is significantly broader than when it initially appeared in literature in the 1980s. Since the turn of this millennium, significant advances in the broader field of biology have provided scientists with unprecedented capabilities for both basic and applied study, and synthetic biology is no exception. The success of mechanical or electrical engineering fascinates synthetic biologists, who aim to engineer living systems to produce desired products. In this respect, the prerequisite for reprogramming the cell as the motherboard is comprehending the fundamental principles or logics of the cell [1]. Through the development of recombinant DNA manipulation techniques, such as BioBricks [2], MAGE [3], and CRISPR-Cas [4, 5], molecular tools have enabled synthetic biologists to modify the genetic network of a host cell to steer its metabolism toward desired chemicals with unprecedented flexibility. Due to the complexity of the cellular environment, it was still difficult to rewire the genetic network with precisely expected functionalities despite the significant progress made in standardizing synthetic parts [6]. While not necessary utilizing the entire cell, pure enzymes and fractionated cell extracts have been employed to study biological reactions and biomanufacturing applications [7]. "Bottom-up" and "top-down" are the two basic techniques of contemporary synthetic biology, which are complementary approaches. The top-down approach seeks to investigate and manipulate existing cellular systems by removing redundant parts or replacing them with synthetic parts and assessing the resulting effects. The bottom-up approach starts with simple building blocks, which are then combined to form a functional cellular system [8]. Long before synthetic biology emerged, similar approaches were used in the field of origins of life to establish a plausible route from non-living matter to a living system, starting with simple elements and inorganic molecules. Today, following the same bottom-up approach, it is possible to envision the task of building a cell from scratch using a wide variety of building blocks.

Early work has demonstrated success in the construction of biological mimicry systems, often known as "protocells" or "primitive cells," with molecules that self-assemble from detergents, surfactants, and fatty acids [9, 10]. Later on, a biomimetic vesicular bilayer system comprised of either synthetic polymers or phospholipids was utilized as the standard synthetic cell model. Not necessarily limited to proteins, numerous catalytic molecules were utilized for their catalytic functions in the construction of these protocell models [11–13]. However, as a result of natural selection throughout the development of living systems, proteins would be the optimal candidate to implement complicated cellular properties in bottom-up synthetic biology [14]. In accordance with the fundamental principles of the Central Dogma, the orchestra of thousands of distinct proteins was ordered by their spatio-temporal expression within the cell. Implementing the gene expression cell-free protein synthesis system within a cellular mimicry system provides the advantage of fulfilling multimer roles beyond simple protocells, according to this perspective [15–17]. The development of cell-free protein synthesis and other molecular tools, such as biomimetic materials [18] and microfabrication for compartmentalization [18], as well as gene expression regulators [19], has accelerated the pace of research into the construction of synthetic minimal cell systems with multiple cellular functions (see Fig. 1). In this review, we focus on the application of cell-free protein synthesis to bottom-up synthetic biology for the development of synthetic cell systems. We addressed the design principles for individual synthetic models using recently reported examples. Finally, we outline existing obstacles and future directions for achieving the ultimate objective of creating a minimum synthetic cell system.

2 Bottom-Up Meets the Top-Down-Development of CFPS System

Long before the well-recognized appearance of the CFPS system utilized by Nirenberg and Matthaei to decipher genetic codes in the 1960s [20], whole cell extracts were already used as important tools to investigate biological reactions. As the first example, back to 1989, Eduard Bucher described the first cell-free fermentation using yeast lysate to convert sugar to alcohol and was awarded the Nobel Prize for chemistry in 1907 [21]. Following the same approach, Harden and Euler-Chelpin received the Nobel Prize in 1929 for their work in discovering the key enzyme and the coenzyme during sugar fermentation [22]. In association with the work of Otto

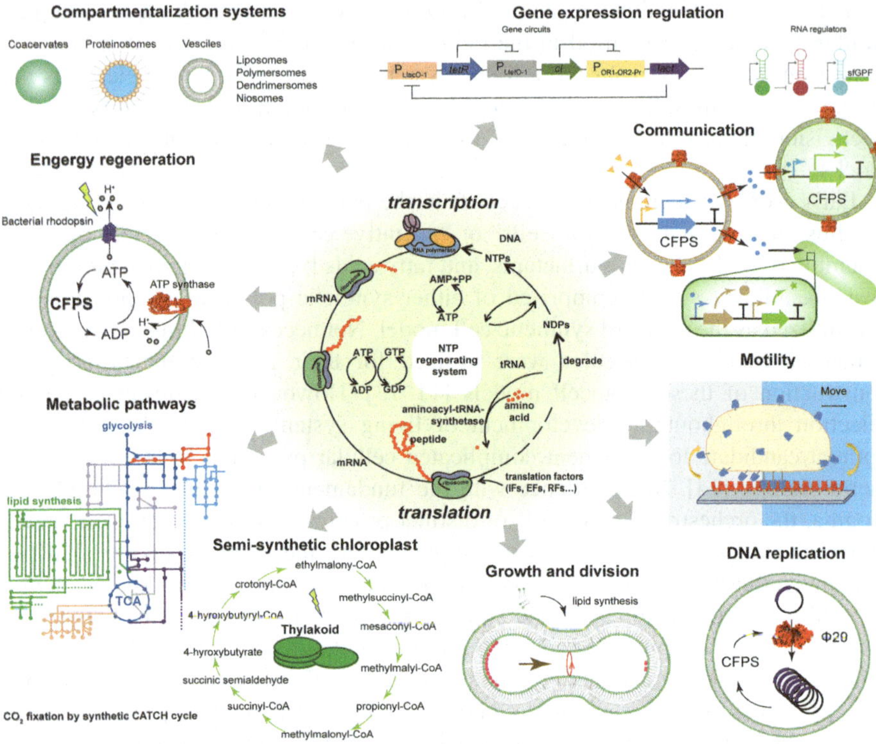

Fig. 1 Bottom-up construction of synthetic cell systems using the CFPS system. Individual modules were developed including compartmentalization, gene expression regulation, energy regeneration, metabolic pathways, semisynthetic chloroplast, growth and division, communication, motility, and self-enhanced DNA replication. Parts of this figure were adapted from Ref. [15] with permission from Copyright 2017, Elsevier B.V.

Meyerhof, who was able to reconstruct in vitro the chain reaction that leads from glycogen to lactic acid, the most fundamental cell metabolic pathway was identified [23]. On this basis, enzyme cascades were stepwise elucidated until Crick first described the central dogma [24] and later the work by Nirenberg and Matthaei, who used the *E. coli* cell lysate to decode the link between the gene and protein. Since then, the cell-free protein synthesis system based on fractionated cell lysate has been widely used as a molecular tool to study many biochemical reactions [25].

Although as a recombinant protein production method, the CFPS technique was noticed much later after the dramatic improvement in productivity from different groups, which included extraction preparation, energy regeneration, buffer optimization, and reaction configurations [26]. Differently from extract-based CFPS system, in 2001, Ueda group described a completely reconstituted cell-free protein synthesis system with purified recombinant proteins (PURE), including T7 RNA polymerase, 20 aminoacyl-tRNA synthetases, translation and ribosome-recycling factors, and ribosomes [27]. At this point, both top-down (cell lysate) and bottom-up (purified elements) methods were used to actualize the basic dogma's core process. However, the initial PURE system was less productive than the lysate-based CFPS approach. By fine-tuning the concentration of specific enzymes and a cytomimetic buffer system, the most recent PURE system was able to achieve a yield comparable to that of the extract-based CFPS system through systematic optimization [28, 29]. Furthermore, with the advent of genome editing techniques, host cells were modified to carry specific functions, which were inherited by the resulting extracts, such as implementing post-translational modifications [30, 31], incorporating noncanonical amino acids, and tuning metabolic pathways by eliminating nonessential enzymes. Such tactics may also be applied using pure enzymes; however, an optimization procedure may be necessary to attain the desired efficiency. The ribosome, the final fractionated component of the PURE system, was recently created using a semisynthetic technique using integrated synthesis, assembly and translation (iSAT), as reported by Jewett's group [32]. Using the iSAT technique, they have demonstrated the creation of functional 30S and 50S subunits, as well as tethered ribosomes [33]. In addition, individual overexpressed ribosomal proteins that make up the 30 s and 50 s subunits can be combined with native rRNAs into a functional ribosome [34–37]. Moreover, this iSAT technique may be confirmed within a vesicle with a double emulsion template [38]. However, despite the successful assembly of functional ribosomes with rRNA produced in vitro for 30S subunits, post-translational changes of 23S rRNA [37] may be required to generate fully reconstituted ribosomes; these alterations must be explored in more depth.

3 Compartmentalization

The initial step in building a synthetic cell was the construction of a physical border to separate the interior space from the exterior world [39]. Various compartmentalization techniques have been developed, including water in oil emulsion droplets [40, 41], unilamellar vesicles (liposomes [42], polymersomes [43], or their hybrids [44]), proteinosomes [45], hydrogel [46], and coacervates [47]. Due to the ease of manipulation provided by agitation or vortexing [48], emulsion droplets were widely used to encapsulate the CFPS system to examine the principles of gene expression such as expression kinetics and macromolecular crowding effects [49, 50]. In addition, the inclusion of surfactants or lipid molecules has stabilized the produced water in oil droplets, thereby extending their lifetime [42, 51]. However, due to the non-natural oil phase outside monolayered droplets, it was not possible to explore characteristics that require the natural two-aqueous environment of a live cell. Instead, phospholipids or structurally comparable synthetic polymers with a hydrophobic tail and hydrophilic head could produce unilamellar vesicles, which are structurally similar to natural cell membranes. Their diameters could be adjusted between 20 nm and a few micrometers using various manufacturing techniques. Nonetheless, cell-sized giant unilamellar vesicles (GUVs) larger than 1 μm were the optimum size for constructing synthetic cell systems, not only because of imaging, but also because of their more homogeneous encapsulation efficiency when compared to smaller vesicles [52]. Despite the structural similarities of phospholipid vesicles to natural cell membranes, synthetic polymer vesicles, however, such as di-, tri-, or grafted block copolymers, possessed superior stability, tunable permeability, and surface features [53]. Nonetheless, synthetic polymer membranes are often thicker and less fluid than lipid membranes, which may impact the integration of integral membrane proteins. To harness the advantage of phospholipid and synthetic polymer membranes, hybrid vesicles were generated with the aim of having the advantage of both materials while compensating for the limitation at the same time [44, 54]. However, it should be noted that separated membranes might form instead of homogeneous bilayers [55]. Alternative compartmentalization strategies, such as proteinosomes and membraneless compartments (hydrogel and coacervates), were verified to successfully encapsulate enzyme-catalyzed processes and CFPS reactions in addition to bilayer-forming compartments [56–59]. See Table 1.

Methods to create unilamellar vesicles can be categorized as solvent-free or solvent displacement [52]. The main processes of solvent-free approaches included creating a thin film of amphiphiles and rehydration in an aqueous solution [71]. The thin dry lipid film on the solid support begins to swell when an aqueous solution is added, eventually forming vesicles as it leaves the surface. This procedure can be aided or sped up by stirring or sonicating. Such rehydration methods usually resulted in the formation of polydispersed multilamellar vesicles [72]. Further extrusion over a fixed-diameter polycarbonate membrane or extended sonication could generate unilamellar vesicles with a more uniform size distribution. Depending on the size of the polycarbonate membrane, the diameter of the resulting vesicles can be adjusted; longer sonication will produce small unilamellar vesicles (diameter less than 100 nM) [73]. Developed from direct rehydration, assisted rehydration methods were developed using electroformation or gel-assisted rehydration, which could form GUVs of cellular size directly [72]. However, the relatively low encapsulation efficiency of macromolecule-rich CFPS systems is a factor to consider when applying the rehydration method to generate vesicles.

Emulsion transfer, also known as double emulsion production of water-in-oil-in-water vesicles, is a typical technique created from emulsion droplets. By passing through another lipid monolayer from the oil phase into the aqueous phase via density difference or centrifugation, the preformed monolayer emulsion droplets could be transformed into a bilayer. Microfluidic chips with successively aligned water–oil and oil–water monolayers could be utilized to further customize this procedure. By injection flow, the solution to be encapsulated would pass through the first monolayer, generating monolayer droplets, and the second interface, forming double emulsion droplets [74]. With uniformly distributed vesicle sizes, the double emulsion technique maintained a high encapsulation efficiency. However, the leftover oil in the lumen of the bilayer interferes with the integration of membrane proteins and the often-underestimated permeability of tiny molecules. To eliminate the leftover oil phase, many techniques were devised, including solvent evaporation, the introduction of additives such as ethanol [75] or surfactant [76] in

Table 1 Summary of compartmentalization

Types	Materials	Methods	Reference
Liposomes	Natural phospholipids or the natural mixture	Solvent-free methods (e.g., film hydration, electroformation and gel-assisted hydration)	[53, 60, 61]
		Solvent-assisted methods (e.g., solvent injection, emulsion phase transfer, microfluidic jetting method, droplet-shooting and size-filtration approach)	[53, 61]
Polymersomes	Amphiphilic polymers	The above methods also apply to the formation of polymers.	[53, 61]
		Polymerization-induced self-assembly (PISA) (a solvent-free method)	[53]
Proteinosomes	Amphiphilic proteins or protein-polymer conjugates	Self-assembly, layer-by-layer technique, W/O interfacial self-assembly method, swelling or extrusion method	[62]
Encapsulins	Capsid-forming proteins	Self-assembly of proteins	[63]
Dendrimersomes	Synthetic amphiphilic Janus dendrimers	Self-assembly in water (e.g., solvent injection and film hydration)	[64]
Niosomes	Mixture of single-alkyl-chain, non-ionic amphipathic surfactant that contains cholesterol and some quantity of ionic surfactant	Similar to the preparation of liposomes (e.g., solvent injection, hand shaking, sonication, reverse-phase evaporation, micro fluidization)	[65, 66]
Coacervates	Polymer-rich mixtures of multivalent, oppositely charged molecules	Liquid–liquid phase separation	[67–69]
Hydrogels	Hydrophilic polymers	Gelation	[70]

the outer aqueous phase, and the substitution of hexane chloroform with octanol [76, 77]. In any case, total solvent removal from vesicles remains a formidable obstacle that will take additional work to resolve [53].

4 Gene Expression Regulation-Introducing Genetic Circuits

If we consider a cellular system as a modern factory, then the spatiotemporal control of the expression of thousands of proteins with varying structures and functions would require immensely complex equipment. Inspired by systematic investigation of gene expression at both the transcriptional and translational levels in vivo, validation of these regulatory elements in vivo began relatively late due to the advanced development of CFPS systems aimed at higher levels of protein expression, often hijacking the most efficient and robust transcription system used in vivo [78]. For example, transcription based on T7 phage polymerase was widely used in the creation of transcription machinery, which was the most robust transcription system in vivo and often led to high background leakage expression [79]. Nevertheless, a systematic study based on the core RNA polymerase from *E.coli* was conducted, validating the regulatory effect on gene expression with different promoters, transcriptional factors (Sigma factors from *E.coli*), RNA polymerases, and repressors [80, 81]. This toolbox allowed the design and testing of various circuit motifs, such as multiple-stage cascades, an AND gate, negative and positive feedback loops, transcriptional cascades with a protein-regulated incoherent feedforward loop, and in vitro ring oscillators [82–89]. In addition, Swank et al. extended the repressing of transcription by introducing an interaction domain to the zinc-finger repressor, which could exhibit a cooperativity on the expression of the reporter gene. By following these design principles, NAND, AND, and OR logic gates can be implemented [90].

Regulating molecules are not limited to proteins, but can also be composed of RNAs. Takahashi [86] was able to effectively establish a transcriptional cascade by employing RNA transcriptional attenuators (pT181 and its variations) as the central regulator. In addition to regulating gene expression at the transcriptional level, RNAs can also exert their influence at the translational level. Changing ribosomal binding sites [91] was a common method for adjusting translation levels since it affected the rate at which ribosomes bound to their respective targets. Riboswitches (small interfering RNAs and RNA thermometers) [89, 92, 93] and ribozymes (RNA molecules that catalyze translation) [94, 95] are two examples of noncoding RNAs that have recently been discovered to function as regulatory elements in relation to translation in order to fine-tune the expression of a given gene. Riboswitches in the 5′-untranslated region (UTR) of mRNA, for example, can regulate upstream gene expression in response to ligand binding at the mRNA level [96, 97]. Lastly, the boundary for tweaking the desired gene expression network should be defined by the load or resource capacity of the system, which includes transcription, translation, and the energy supporting system.

5 Energy Regeneration and Metabolism

Energy regeneration is a crucial characteristic of a living cell, enabling various biochemical reactions which are a fundamental requirement for a synthetic cell. In a living cell, nutrients are broken down (catabolism) through a set of orchestrated biochemical reactions to release energy, which is then stored in ATP, the universal energy currency of cells. In aerobic cells, this coupling process typically comprises three parts: glycolysis, the Krebs cycle, and oxidative phosphorylation. Glucose

[98–102] and other key compounds within the glycolysis pathways [103–107] are used to efficiently fuel CFPS systems by activating either endogenous or extra kinases, which can support the running of such reactions for several hours in batch mode and up to several days with a continuous supply of energy precursors [42, 107]. However, the continuous supply of such energy precursors in the current artificial cell model is challenging and would require the functional integration of transporters and antiporters to maintain a homeostatic environment inside the compartment. On the other hand, ATP can be robustly generated via the conserved membrane protein complex ATP synthase, which utilizes the proton gradient as the driving force. Such electrochemical gradients can be established via oxidative phosphorylation from heterotrophic organisms, or photosynthesis, light-driven proton-pump proteins (rhodopsins) from phototrophic organisms [108]. As the most straightforward energy harvesting approach, coupling the ATP synthase with the proton pump establishes direct energy recycling by harnessing light. Coupling the light-activated bacterial proton-pump rhodopsin with the ATP synthase to generate ATP triggered by light has been known for some time [109–111]. However, the reverse activity of ATP synthase [112] rapidly consuming the newly produced ATP molecules limited its ability as a continuous energy source to support reaction such as CFPS. As shown by previous work by Bald et al. and recent work by Berhanu et al. [113, 114], this challenge can be addressed by specifically blocking the ATPase activity of ATP synthase with sodium azide. The introduction of sodium azide has greatly increased the lifespan of ATP synthesis, allowing it to be sufficient to energize the PURE system for protein expression.

Furthermore, by synthesizing both bacterial rhodopsin and parts of the subunits of the ATP synthase complex, it was possible to build a light-driven synthetic energy regeneration system with positive feedback [114]. Taking this step further, Lee et al. [115] constructed an artificial organelle by simultaneously incorporating proteorhodopsin and photosynthesis complex II, which was able to function as an artificial mitochondrion for ATP generation or a chloroplast for CO_2 fixation.

In the work by Biner [116], the reconstitution of the respiratory chain was demonstrated by reconstituting mitochondrial complex I, an alternative oxidase and ATP synthase, which can be driven by different fuels through NAD+/NADH cycling to support the protein expression of the CFPS system within GUVs. Moreover, the breakdown of arginine, along with a coupled transporter and antiporter, could also provide a continuous energy supply for anabolism [117]. The recent demonstration of Erb's group [118] has allowed others to think of potentials to create a complete synthetic chloroplast system. By coencapsulating the purified thylakoids and a synthetic pathway – the crotonyl-CoA/ethylmalonyl-CoA/hydroxybutyryl-CoA (CETCH) cycle, they could demonstrate the CO_2 fixation through light. However, creating a complete reconstituted photosynthesis system would be the next logical, yet challenging step for a fully synthetic chloroplast.

6 Growth and Division

Although less studied, the ability to divide is an essential characteristic of any living entity and is indispensable for the maintenance and evolution of a living system [119]. Compared to other aspects such as compartmentalization, protein expression and regulation, energy regeneration, or metabolism, the ability to divide appears more straightforward, as it can be triggered merely by mechanical forces. The division of a natural cell is a process that involves a precisely orchestrated group of different molecules. Early work by fatty acid droplets has successfully proven that surface and volume expansion can induce the division of synthetic vesicles [42, 120]. In addition, membraneless compartments such as phase separation can divide through volume growth and energy dissipation, as demonstrated by Zwicker et al. [121–123]. Lipid membranes are soft objects, which can be easily deformed by surface expansion, but it is difficult to achieve complete fission [119]. Several examples have shown the growth of different microcompartments without enzymes [124]; however, we envisioned that de novo synthesis of lipid molecules would be vital for autonomous self-reproduction. Previous studies showed the successful synthesis of PE and PG within the synthetic cell by activating the synthesis of correlated enzymes by the encapsulated CFPS system [125, 126]. Back to the cell, in most single cell systems, mitosis was driven by active self-assembled proteins, which formed ring structures such as the actomyosin ring formation of eukaryotic cells or the FtsZ ring of prokaryotic cells [127–129]. However, the detailed mechanism by which these ring structures developed into a complete fission is still unclear. The constriction of actomyosin rings in the eukaryotic system was originally thought to proceed, following the line as known from sarcomeres, with actin and myosin filaments moving in opposite directions, and therefore restricting the radius of the ring, which was still not confirmed. Instead, myosin appears to cause actin filament contraction by destabilizing them [130]. However, as the trigger for constriction, it is not yet unknown whether this is sufficient for ring contraction to the point of membrane fission [131]. Recent advances in reconstitution of the actomyosin system have successfully demonstrated the formation of actin meshwork, actomyosin ring, actin cortex membrane anchoring, and dynamic deformed membranes of the actin cortex [132]. However, challenges remain in many detailed processes on how to transform deformed vesicles via the dynamics of actomyosin structures to the ultimate membrane fission. In prokaryotes, treadmilling of the self-assembled tubulin protein FtsZ was deemed to generate the force for a large-scale membrane bending force with GTP hydrolysis. However, a recent in vivo study pointed to an indirect effect of this treadmilling along the cell periphery, leading to the recruitment of enzymes responsible for cell wall synthesis in the middle of the cell [133–135]. Instead of focusing on the ring structure, many studies have been done to elucidate geometry sensing machinery – how to correctly position the Z ring at midcell. The MinDE protein system was intensively studied both in vivo and in vitro [136, 137]. Recently, a fully reconstituted system consisting of Min C, D, E, FtsA, and FtsZ has been described by de novo synthesis by a PURE system within a

GUV [138]. They managed to show that a ring-induced shape transformation of the formed GUVs was observed, which might be the leading forces to trigger further deformation until the GUVs are divided. However, continuous efforts are still required to achieve an autonomous, self-regulated division in any synthetic cell systems.

The precondition for such growth-induced division would require the coupling of the growing surface, which ultimately would require the continuous production of phospholipids in the case of a phospholipid vesicle. In addition to the continuous supply of lipid molecules from the solution, the de novo synthesis of phospholipids could provide valuable insights toward the expansion of commonly used phospholipid vesicles via a CFPS system.

7 Communication and Motility

Cells in their natural state use a process called quorum sensing to receive information from their immediate surroundings and from other cells. This interaction, when coupled with the CFPS system, could be used to activate responses in both natural and synthetic cells [139–142]. Efforts have been made by synthetic biologists to create orthogonal communication channels, which could aid in the engineering of microorganisms without the tedious in vivo work. Although, it is not easy to keep several activators from interfering with one another. In a recent study, Halleran and Murray [143] demonstrated that the CFPS method could be a quick technique to validate crosstalk before engineering work in vivo. Adamala et al. [144] further demonstrated that the synthesis of the pair of SNARE proteins led to the selective fusion of synthetic cells. As demonstrated by different groups, IPTG and alpha hemolysin have been successfully used as sender and receiver for communication between synthetic and natural cells [88, 145–147].

Although even less investigated, cellular motility is a vital process essential for embryonic development, wound healing, immune responses, and tissue developments. An early concept based on enzyme-catalyzed repulsion force generation was reported. However, a recent synthetic cell with reversible light-activated binding protein pairs (iLID and Micro) was able to achieve a light-direct movement of GUVs on a solid support [148]. However, substantial efforts based on the cytoskeletal process driven by ATP hydrolysis and the discovery of symmetry breaking mechanisms that overlap with a portion of cell division were continuously developed, covered in detail elsewhere [149–154].

8 Integration of Individual Modules

Technical advances have enabled synthetic biologists to envision complex systems with the integration of individually developed modules, using CFPS systems as central platforms for gene expression and regulatory toolboxes. In the context of constructing a minimal synthetic cell system, autonomous self-replication has been demonstrated in some impressive recent cases. For example, the Danelon group showcased a self-replicating artificial cell, which was able to replicate itself by coencapsulating the DNA template encoded with the phi29 polymerase through the PURE system [155]. In this case, the DNA of phi29 was replicated when the coding proteins were expressed by the PURE system. Furthermore, Libicher et al. [156] were able to replicate a 116 kb multipartite genome (distributed in 11 plasmids) via the PURE system – a size that matches the minimum genome edited from the top-down approach. Furthermore, individual genes within such plasmids can be synthesized, further enhancing the productivity of the established system. Lastly, recent advances in ribosome biogenesis, including iSAT and de novo synthesis of ribosome proteins, provide a blueprint for a long-lived and continuously self-replicating synthetic cell system [32, 35, 36, 38].

9 Challenges and Perspectives

Although there have been many efforts to boost the efficiency of CFPS systems, there is still a lot of potential to improve the yield and bring it closer to the efficiency of real cells. The average ribosome in *E. coli* can catalyze around 55,000 peptide bonds in a 20 min doubling time, which is almost an order of magnitude higher than the current CFPS system [15, 35]. Furthermore, the challenge of encapsulating the different components of synthetic cells, which contain high-viscosity macromolecules, is still present. To integrate self-sustaining and long-lived synthetic cells, the vesicles need to be stable and the inner environment homeostatic and be able to exchange small molecules, which can be achieved by integrating functional membrane proteins. CFPS systems offer a quick and versatile platform for protein engineering, enabling directed evolution [157–159], site-specific incorporation of noncanonical amino acids [160–162], and rational design [163]. Artificial intelligence has been rapidly evolving, which will help bottom-up synthetic biology in designing proteins and systems to coordinate the integration of different synthetic modules. CFPS systems have already been utilized for product-driven research, such as metabolic engineering [163, 164], biomedical engineering [165], and in vitro diagnostics [166, 167], which have to gain from the progress made in addressing fundamental questions related to minimal synthetic cell systems. Therefore, we believe that the broader field of synthetic biology and biotechnology as a whole would benefit greatly from such advances.

References

1. Forster AC, Church GM (2007) Synthetic biology projects in vitro. Genome Res 17:1–6
2. Shetty RP, Endy D, Knight Jr TF (2008) Engineering BioBrick vectors from BioBrick parts. J Biol Eng 2:5

3. Wang HH, Isaacs FJ, Carr PA, Sun ZZ, Xu G, Forest CR, Church GM (2009) Programming cells by multiplex genome engineering and accelerated evolution. Nature 460:894–898
4. Cong L, Ran FA, Cox D, Lin S, Barretto R, Habib N, Hsu PD, Wu X, Jiang W, Marraffini LA, Zhang F (2013) Multiplex genome engineering using CRISPR/Cas systems. Science 339:819–823
5. Doudna JA, Charpentier E (2014) Genome editing. The new frontier of genome engineering with CRISPR-Cas9. Science 346:1258096
6. Wu G, Yan Q, Jones JA, Tang YJ, Fong SS, Koffas MAG (2016) Metabolic burden: cornerstones in synthetic biology and metabolic engineering applications. Trends Biotechnol 34:652–664
7. Hodgman CE, Jewett MC (2012) Cell-free synthetic biology: thinking outside the cell. Metab Eng 14:261–269
8. Forster AC, Church GM (2006) Towards synthesis of a minimal cell. Mol Syst Biol 2:45
9. Luisi PL, Varela FJ (1989) Self-replicating micelles – a chemical version of a minimal autopoietic system. Orig Life Evol Biosph 19:633–643
10. Stano P, Luisi PL (2010) Achievements and open questions in the self-reproduction of vesicles and synthetic minimal cells. Chem Commun (Camb) 46:3639–3653
11. Rasmussen S, Chen L, Nilsson M, Abe S (2003) Bridging nonliving and living matter. Artif Life 9:269–316
12. Hanczyc MM, Fujikawa SM, Szostak JW (2003) Experimental models of primitive cellular compartments: encapsulation, growth, and division. Science 302:618–622
13. Mansy SS, Schrum JP, Krishnamurthy M, Tobe S, Treco DA, Szostak JW (2008) Template-directed synthesis of a genetic polymer in a model protocell. Nature 454:122–125
14. Schwille P, Spatz J, Landfester K, Bodenschatz E, Herminghaus S, Sourjik V, Erb TJ, Bastiaens P, Lipowsky R, Hyman A, Dabrock P, Baret JC, Vidakovic-Koch T, Bieling P, Dimova R, Mutschler H, Robinson T, Tang TD, Wegner S, Sundmacher K (2018) MaxSynBio: avenues towards creating cells from the bottom up. Angew Chem Int Ed Engl 57:13382–13392
15. Jia H, Heymann M, Bernhard F, Schwille P, Kai L (2017) Cell-free protein synthesis in micro compartments: building a minimal cell from biobricks. N Biotechnol 39:199–205
16. Cho E, Lu Y (2020) Compartmentalizing cell-free systems: toward creating life-like artificial cells and beyond. ACS Synth Biol 9:2881–2901
17. Ivanov I, Castellanos SL, Balasbas III S, Otrin L, Marusic N, Vidakovic-Koch T, Sundmacher K (2021) Bottom-up synthesis of artificial cells: recent highlights and future challenges. Annu Rev Chem Biomol Eng 12:287–308
18. Fu Z, Ochsner MA, De Hoog HP, Tomczak N, Nallani M (2011) Multicompartmentalized polymersomes for selective encapsulation of biomacromolecules. Chem Commun (Camb) 47: 2862–2864
19. Garamella J, Marshall R, Rustad M, Noireaux V (2016) The all E. coli TX-TL toolbox 2.0: a platform for cell-free synthetic biology. ACS Synth Biol 5:344–355
20. Nirenberg MW, Matthaei JH (1961) The dependence of cell-free protein synthesis in E. coli upon naturally occurring or synthetic polyribonucleotides. Proc Natl Acad Sci U S A 47:1588–1602
21. Buchner E (1907) Nobel lecture: cell-free fermentation. In: Nobelprize.org (ed) Nobel Media AB; 2013
22. Meyerhof O (1945) The origin of the reaction of harden and young in cell-free alcoholic fermentation. J Biol Chem 157:105–119
23. Meyerhof OF (1923) Energy conversions in muscle. Nobel Lecture. 11 Dec 1923, https://old.nobelprize.org/nobel_prizes/medicine/laureates/1922/meyerhof-lecture.html
24. Crick F (1970) Central dogma of molecular biology. Nature 227:561–563
25. Zubay G (1973) In vitro synthesis of protein in microbial systems. Annu Rev Genet 7:267–287
26. Carlson ED, Gan R, Hodgman CE, Jewett MC (2012) Cell-free protein synthesis: applications come of age. Biotechnol Adv 30:1185–1194

27. Shimizu Y, Inoue A, Tomari Y, Suzuki T, Yokogawa T, Nishikawa K, Ueda T (2001) Cell-free translation reconstituted with purified components. Nat Biotechnol 19:751–755
28. Shimizu Y, Kanamori T, Ueda T (2005) Protein synthesis by pure translation systems. Methods 36:299–304
29. Kazuta Y, Matsuura T, Ichihashi N, Yomo T (2014) Synthesis of milligram quantities of proteins using a reconstituted in vitro protein synthesis system. J Biosci Bioeng 118:554–557
30. Kai L, Sonal HT, Schwille P (2023) Reconstitution of a reversible membrane switch via prenylation by one-pot cell-free expression. ACS Synth Biol 12:108–119
31. Jaroentomeechai T, Stark JC, Natarajan A, Glasscock CJ, Yates LE, Hsu KJ, Mrksich M, Jewett MC, Delisa MP (2018) Single-pot glycoprotein biosynthesis using a cell-free transcription-translation system enriched with glycosylation machinery. Nat Commun 9:2686
32. Jewett MC, Fritz BR, Timmerman LE, Church GM (2013) In vitro integration of ribosomal RNA synthesis, ribosome assembly, and translation. Mol Syst Biol 9:678
33. Orelle C, Carlson ED, Szal T, Florin T, Jewett MC, Mankin AS (2015) Protein synthesis by ribosomes with tethered subunits. Nature 524:119–124
34. Traub P, Nomura M (1968) Structure and function of E. coli ribosomes. V. Reconstitution of functionally active 30S ribosomal particles from RNA and proteins. Proc Natl Acad Sci U S A 59:777–784
35. Li J, Haas W, Jackson K, Kuru E, Jewett MC, Fan ZH, Gygi S, Church GM (2017) Cogenerating synthetic parts toward a self-replicating system. ACS Synth Biol 6:1327–1336
36. Aoyama R, Masuda K, Shimojo M, Kanamori T, Ueda T, Shimizu Y (2022) In vitro reconstitution of the Escherichia coli 70S ribosome with a full set of recombinant ribosomal proteins. J Biochem 171:227–237
37. Semrad K, Green R (2002) Osmolytes stimulate the reconstitution of functional 50S ribosomes from in vitro transcripts of Escherichia coli 23S rRNA. RNA 8:401–411
38. Caschera F, Lee JW, Ho KK, Liu AP, Jewett MC (2016) Cell-free compartmentalized protein synthesis inside double emulsion templated liposomes with in vitro synthesized and assembled ribosomes. Chem Commun (Camb) 52:5467–5469
39. Ai Y, Xie R, Xiong J, Liang Q (2020) Microfluidics for biosynthesizing: from droplets and vesicles to artificial cells. Small 16:e1903940
40. Dittrich PS, Jahnz M, Schwille P (2005) A new embedded process for compartmentalized cell-free protein expression and on-line detection in microfluidic devices. Chembiochem 6:811–814
41. Torre P, Keating CD, Mansy SS (2014) Multiphase water-in-oil emulsion droplets for cell-free transcription-translation. Langmuir 30:5695–5699
42. Noireaux V, Libchaber A (2004) A vesicle bioreactor as a step toward an artificial cell assembly. Proc Natl Acad Sci U S A 101:17669–17674
43. Nallani M, Andreasson-Ochsner M, Tan CW, Sinner EK, Wisantoso Y, Geifman-Shochat S, Hunziker W (2011) Proteopolymersomes: in vitro production of a membrane protein in polymersome membranes. Biointerphases 6:153–157
44. Jacobs ML, Kamat NP (2022) Cell-free membrane protein expression into hybrid lipid/polymer vesicles. Methods Mol Biol 2433:257–271
45. Huang X, Patil AJ, Li M, Mann S (2014) Design and construction of higher-order structure and function in proteinosome-based protocells. J Am Chem Soc 136:9225–9234
46. Whitfield CJ, Banks AM, Dura G, Love J, Fieldsend JE, Goodchild SA, Fulton DA, Howard TP (2020) Cell-free protein synthesis in hydrogel materials. Chem Commun (Camb) 56:7108–7111
47. Tang T-YD, Van Swaay D, Demello A, Anderson JR, Mann S (2015) In vitro gene expression within membrane-free coacervate protocells. Chem Commun (Camb) 51:11429–11432
48. Tawfik DS, Griffiths AD (1998) Man-made cell-like compartments for molecular evolution. Nat Biotechnol 16:652–656

49. Hansen MM, Meijer LH, Spruijt E, Maas RJ, Rosquelles MV, Groen J, Heus HA, Huck WT (2016) Macromolecular crowding creates heterogeneous environments of gene expression in picolitre droplets. Nat Nanotechnol 11:191–197
50. Hansen MM, Paffenholz S, Foschepoth D, Heus HA, Thiele J, Huck WT (2016) Cell-like nanostructured environments alter diffusion and reaction kinetics in cell-free gene expression. Chembiochem 17:228–232
51. Loakes D, Holliger P (2009) Darwinian chemistry: towards the synthesis of a simple cell. Mol Biosyst 5:686–694
52. Walde P, Cosentino K, Engel H, Stano P (2010) Giant vesicles: preparations and applications. Chembiochem 11:848–865
53. Rideau E, Dimova R, Schwille P, Wurm FR, Landfester K (2018) Liposomes and polymersomes: a comparative review towards cell mimicking. Chem Soc Rev 47:8572–8610
54. Elani Y (2021) Interfacing living and synthetic cells as an emerging frontier in synthetic biology. Angew Chem Int Ed Engl 60:5602–5611
55. Le Meins JF, Schatz C, Lecommandoux S, Sandre O (2013) Hybrid polymer/lipid vesicles: state of the art and future perspectives. Mater Today 16:397–402
56. Thiele J, Ma Y, Foschepoth D, Hansen MM, Steffen C, Heus HA, Huck WT (2014) DNA-functionalized hydrogels for confined membrane-free in vitro transcription/translation. Lab Chip 14:2651–2656
57. Zhou X, Wu H, Cui M, Lai SN, Zheng B (2018) Long-lived protein expression in hydrogel particles: towards artificial cells. Chem Sci 9:4275–4279
58. Lai SN, Zhou X, Ouyang X, Zhou H, Liang Y, Xia J, Zheng B (2020) Artificial cells capable of long-lived protein synthesis by using aptamer grafted polymer hydrogel. ACS Synth Biol 9: 76–83
59. Schoonen L, Van Hest JC (2016) Compartmentalization approaches in soft matter science: from nanoreactor development to organelle mimics. Adv Mater 28:1109–1128
60. Szoka Jr F, Papahadjopoulos D (1980) Comparative properties and methods of preparation of lipid vesicles (liposomes). Annu Rev Biophys Bioeng 9:467–508
61. Seo H, Lee H (2021) Recent developments in microfluidic synthesis of artificial cell-like polymersomes and liposomes for functional bioreactors. Biomicrofluidics 15:021301
62. Wang XL, Liu XM, Huang X (2020) Bioinspired protein-based assembling: toward advanced life-like behaviors. Adv Mater 32:e2001436
63. Sutter M, Boehringer D, Gutmann S, Günther S, Prangishvili D, Loessner MJ, Stetter KO, Weber-Ban E, Ban N (2008) Structural basis of enzyme encapsulation into a bacterial nanocompartment. Nat Struct Mol Biol 15:939–947
64. Percec V, Wilson DA, Leowanawat P, Wilson CJ, Hughes AD, Kaucher MS, Hammer DA, Levine DH, Kim AJ, Bates FS, Davis KP, Lodge TP, Klein ML, Devane RH, Aqad E, Rosen BM, Argintaru AO, Sienkowska MJ, Rissanen K, Nummelin S, Ropponen J (2010) Self-assembly of Janus dendrimers into uniform dendrimersomes and other complex architectures. Science 328:1009–1014
65. Baillie AJ, Florence AT, Hume LR, Muirhead GT, Rogerson A (1985) The preparation and properties of niosomes--non-ionic surfactant vesicles. J Pharm Pharmacol 37:863–868
66. Umbarkar MG (2021) Niosome as a novel pharmaceutical drug delivery: a brief review highlighting formulation, types, composition and application. Indian J Pharm Educ 55:S11–S28
67. Lu T, Spruijt E (2020) Multiphase complex coacervate droplets. J Am Chem Soc 142:2905–2914
68. Poudyal RR, Pir Cakmak F, Keating CD, Bevilacqua PC (2018) Physical principles and extant biology reveal roles for RNA-containing membraneless compartments in origins of life chemistry. Biochemistry 57:2509–2519
69. Mason AF, Van Hest JCM (2019) Multifaceted cell mimicry in coacervate-based synthetic cells. Emerg Top Life Sci 3:567–571

70. Tan H, Guo S, Dinh ND, Luo R, Jin L, Chen CH (2017) Heterogeneous multi-compartmental hydrogel particles as synthetic cells for incompatible tandem reactions. Nat Commun 8:663
71. Rideau E, Wurm FR, Landfester K (2019) Self-assembly of Giant Unilamellar vesicles by film hydration methodologies. Adv Biosyst 3:e1800324
72. Yuan W, Piao J, Dong Y (2021) Advancements in the preparation methods of artificial cell membranes with lipids. Mater Chem Front 5:5233–5246
73. Woodbury DJ, Richardson ES, Grigg AW, Welling RD, Knudson BH (2006) Reducing liposome size with ultrasound: bimodal size distributions. J Liposome Res 16:57–80
74. Damiati S, Mhanna R, Kodzius R, Ehmoser EK (2018) Cell-free approaches in synthetic biology utilizing microfluidics. Genes (Basel) 9:144
75. Teh SY, Khnouf R, Fan H, Lee AP (2011) Stable, biocompatible lipid vesicle generation by solvent extraction-based droplet microfluidics. Biomicrofluidics 5:44113–4411312
76. Deng NN, Yelleswarapu M, Huck WT (2016) Monodisperse Uni- and Multicompartment Liposomes. J Am Chem Soc 138:7584–7591
77. Deshpande S, Caspi Y, Meijering AE, Dekker C (2016) Octanol-assisted liposome assembly on chip. Nat Commun 7:10447
78. Hasty J, Mcmillen D, Collins JJ (2002) Engineered gene circuits. Nature 420:224–230
79. Wang W, Li Y, Wang Y, Shi C, Li C, Li Q, Linhardt RJ (2018) Bacteriophage T7 transcription system: an enabling tool in synthetic biology. Biotechnol Adv 36:2129–2137
80. Shin J, Noireaux V (2010) Efficient cell-free expression with the endogenous E. Coli RNA polymerase and sigma factor 70. J Biol Eng 4:8
81. Shin J, Noireaux V (2012) An E. coli cell-free expression toolbox: application to synthetic gene circuits and artificial cells. ACS Synth Biol 1:29–41
82. Karzbrun E, Shin J, Bar-Ziv RH, Noireaux V (2011) Coarse-grained dynamics of protein synthesis in a cell-free system. Phys Rev Lett 106:048104
83. Takahashi MK, Hayes CA, Chappell J, Sun ZZ, Murray RM, Noireaux V, Lucks JB (2015) Characterizing and prototyping genetic networks with cell-free transcription-translation reactions. Methods 86:60–72
84. Sun ZZ, Yeung E, Hayes CA, Noireaux V, Murray RM (2014) Linear DNA for rapid prototyping of synthetic biological circuits in an Escherichia coli based TX-TL cell-free system. ACS Synth Biol 3:387–397
85. Siegal-Gaskins D, Tuza ZA, Kim J, Noireaux V, Murray RM (2014) Gene circuit performance characterization and resource usage in a cell-free "breadboard". ACS Synth Biol 3:416–425
86. Takahashi MK, Chappell J, Hayes CA, Sun ZZ, Kim J, Singhal V, Spring KJ, Al-Khabouri S, Fall CP, Noireaux V, Murray RM, Lucks JB (2015) Rapidly characterizing the fast dynamics of RNA genetic circuitry with cell-free transcription-translation (TX-TL) systems. ACS Synth Biol 4:503–515
87. Chappell J, Takahashi MK, Lucks JB (2015) Creating small transcription activating RNAs. Nat Chem Biol 11:214–220
88. Niederholtmeyer H, Sun ZZ, Hori Y, Yeung E, Verpoorte A, Murray RM, Maerkl SJ (2015) Rapid cell-free forward engineering of novel genetic ring oscillators. Elife 4:e09771
89. Sadler FW, Dodevski I, Sarkar CA (2018) RNA thermometers for the PURExpress system. ACS Synth Biol 7:292–296
90. Swank Z, Laohakunakorn N, Maerkl SJ (2019) Cell-free gene-regulatory network engineering with synthetic transcription factors. Proc Natl Acad Sci U S A 116:5892–5901
91. Chizzolini F, Forlin M, Cecchi D, Mansy SS (2014) Gene position more strongly influences cell-free protein expression from operons than T7 transcriptional promoter strength. ACS Synth Biol 3:363–371
92. Debroy S, Gebbie M, Ramesh A, Goodson JR, Cruz MR, Van Hoof A, Winkler WC, Garsin DA (2014) Riboswitches. A riboswitch-containing sRNA controls gene expression by sequestration of a response regulator. Science 345:937–940
93. Jia H, Heymann M, Hartel T, Kai L, Schwille P (2019) Temperature-sensitive protein expression in protocells. Chem Commun (Camb) 55:6421–6424

94. Wieland M, Hartig JS (2008) Artificial riboswitches: synthetic mRNA-based regulators of gene expression. Chembiochem 9:1873–1878
95. Weigand JE, Suess B (2009) Aptamers and riboswitches: perspectives in biotechnology. Appl Microbiol Biotechnol 85:229–236
96. Winkler WC (2005) Riboswitches and the role of noncoding RNAs in bacterial metabolic control. Curr Opin Chem Biol 9:594–602
97. Martini L, Mansy SS (2011) Cell-like systems with riboswitch controlled gene expression. Chem Commun (Camb) 47:10734–10736
98. Calhoun KA, Swartz JR (2005) An economical method for cell-free protein synthesis using glucose and nucleoside monophosphates. Biotechnol Prog 21:1146–1153
99. Calhoun KA, Swartz JR (2005) Energizing cell-free protein synthesis with glucose metabolism. Biotechnol Bioeng 90:606–613
100. Anderson MJ, Stark JC, Hodgman CE, Jewett MC (2015) Energizing eukaryotic cell-free protein synthesis with glucose metabolism. FEBS Lett 589:1723–1727
101. Kim TW, Oh IS, Keum JW, Kwon YC, Byun JY, Lee KH, Choi CY, Kim DM (2007) Prolonged cell-free protein synthesis using dual energy sources: combined use of creatine phosphate and glucose for the efficient supply of ATP and retarded accumulation of phosphate. Biotechnol Bioeng 97:1510–1515
102. Kim H-C, Kim T-W, Kim D-M (2011) Prolonged production of proteins in a cell-free protein synthesis system using polymeric carbohydrates as an energy source. Process Biochem 46: 1366–1369
103. Jewett MC, Swartz JR (2004) Mimicking the Escherichia coli cytoplasmic environment activates long-lived and efficient cell-free protein synthesis. Biotechnol Bioeng 86:19–26
104. Kim DM, Swartz JR (2001) Regeneration of adenosine triphosphate from glycolytic intermediates for cell-free protein synthesis. Biotechnol Bioeng 74:309–316
105. Kim DM, Swartz JR (1999) Prolonging cell-free protein synthesis with a novel ATP regeneration system. Biotechnol Bioeng 66:180–188
106. Kim TW, Keum JW, Oh IS, Choi CY, Kim HC, Kim DM (2007) An economical and highly productive cell-free protein synthesis system utilizing fructose-1,6-bisphosphate as an energy source. J Biotechnol 130:389–393
107. Caschera F, Noireaux V (2014) Synthesis of 2.3 mg/ml of protein with an all Escherichia coli cell-free transcription-translation system. Biochimie 99:162–168
108. Otrin L, Kleineberg C, Caire Da Silva L, Landfester K, Ivanov I, Wang M, Bednarz C, Sundmacher K, Vidakovic-Koch T (2019) Artificial organelles for energy regeneration. Adv Biosyst 3:e1800323
109. Racker E, Stoeckenius W (1974) Reconstitution of purple membrane vesicles catalyzing light-driven proton uptake and adenosine triphosphate formation. J Biol Chem 249:662–663
110. Yoshida M, Sone N, Hirata H, Kagawa Y (1975) ATP synthesis catalyzed by purified DCCD-sensitive ATPase incorporated into reconstituted purple membrane vesicles. Biochem Biophys Res Commun 67:1295–1300
111. Hara KY, Suzuki R, Suzuki T, Yoshida M, Kino K (2011) ATP photosynthetic vesicles for light-driven bioprocesses. Biotechnol Lett 33:1133–1138
112. Tuena Gómez-Puyou MT, Martins OB, Gómez-Puyou A (1988) Synthesis and hydrolysis of ATP by the mitochondrial ATP synthase. Biochem Cell Biol 66:677–682
113. Bald D, Amano T, Muneyuki E, Pitard B, Rigaud JL, Kruip J, Hisabori T, Yoshida M, Shibata M (1998) ATP synthesis by F0F1-ATP synthase independent of noncatalytic nucleotide binding sites and insensitive to azide inhibition. J Biol Chem 273:865–870
114. Berhanu S, Ueda T, Kuruma Y (2019) Artificial photosynthetic cell producing energy for protein synthesis. Nat Commun 10:1325
115. Lee KY, Park SJ, Lee KA, Kim SH, Kim H, Meroz Y, Mahadevan L, Jung KH, Ahn TK, Parker KK, Shin K (2018) Photosynthetic artificial organelles sustain and control ATP-dependent reactions in a protocellular system. Nat Biotechnol 36:530–535

116. Biner O, Fedor JG, Yin Z, Hirst J (2020) Bottom-up construction of a minimal system for cellular respiration and energy regeneration. ACS Synth Biol 9:1450–1459
117. Pols T, Singh S, Deelman-Driessen C, Gaastra BF, Poolman B (2021) Enzymology of the pathway for ATP production by arginine breakdown. FEBS J 288:293–309
118. Miller TE, Beneyton T, Schwander T, Diehl C, Girault M, Mclean R, Chotel T, Claus P, Cortina NS, Baret JC, Erb TJ (2020) Light-powered CO(2) fixation in a chloroplast mimic with natural and synthetic parts. Science 368:649–654
119. Schwille P (2019) Division in synthetic cells. Emerg Top Life Sci 3:551–558
120. Dobereiner HG, Kas J, Noppl D, Sprenger I, Sackmann E (1993) Budding and fission of vesicles. Biophys J 65:1396–1403
121. Zwicker D, Hyman AA, Julicher F (2015) Suppression of Ostwald ripening in active emulsions. Phys Rev E Stat Nonlinear Soft Matter Phys 92:012317
122. Zwicker D, Seyboldt R, Weber CA, Hyman AA, Jülicher F (2016) Growth and division of active droplets provides a model for protocells. Nat Phys 13:408–413
123. Rosowski KA, Sai T, Vidal-Henriquez E, Zwicker D, Style RW, Dufresne ER (2020) Elastic ripening and inhibition of liquid–liquid phase separation. Nat Phys 16:422–425
124. Exterkate M, Driessen AJM (2019) Synthetic minimal cell: self-reproduction of the boundary layer. ACS Omega 4:5293–5303
125. Scott A, Noga MJ, De Graaf P, Westerlaken I, Yildirim E, Danelon C (2016) Cell-free phospholipid biosynthesis by gene-encoded enzymes reconstituted in liposomes. PloS One 11:e0163058
126. Blanken D, Foschepoth D, Serrao AC, Danelon C (2020) Genetically controlled membrane synthesis in liposomes. Nat Commun 11:4317
127. Lutkenhaus J, Addinall SG (1997) Bacterial cell division and the Z ring. Annu Rev Biochem 66:93–116
128. Scholey JM, Brust-Mascher I, Mogilner A (2003) Cell division. Nature 422:746–752
129. Cheffings TH, Burroughs NJ, Balasubramanian MK (2016) Actomyosin ring formation and tension generation in eukaryotic cytokinesis. Curr Biol 26:R719–R737
130. Mendes Pinto I, Rubinstein B, Kucharavy A, Unruh JR, Li R (2012) Actin depolymerization drives actomyosin ring contraction during budding yeast cytokinesis. Dev Cell 22:1247–1260
131. Pollard TD (2017) Nine unanswered questions about cytokinesis. J Cell Biol 216:3007–3016
132. Baldauf L, Van Buren L, Fanalista F, Koenderink GH (2022) Actomyosin-driven division of a synthetic cell. ACS Synth Biol 11:3120–3133
133. Monteiro JM, Pereira AR, Reichmann NT, Saraiva BM, Fernandes PB, Veiga H, Tavares AC, Santos M, Ferreira MT, Macario V, Vannieuwenhze MS, Filipe SR, Pinho MG (2018) Peptidoglycan synthesis drives an FtsZ-treadmilling-independent step of cytokinesis. Nature 554:528–532
134. Yang X, Lyu Z, Miguel A, Mcquillen R, Huang KC, Xiao J (2017) GTPase activity-coupled treadmilling of the bacterial tubulin FtsZ organizes septal cell wall synthesis. Science 355: 744–747
135. Bisson-Filho AW, Hsu YP, Squyres GR, Kuru E, Wu F, Jukes C, Sun Y, Dekker C, Holden S, Vannieuwenhze MS, Brun YV, Garner EC (2017) Treadmilling by FtsZ filaments drives peptidoglycan synthesis and bacterial cell division. Science 355:739–743
136. Loose M, Kruse K, Schwille P (2011) Protein self-organization: lessons from the min system. Annu Rev Biophys 40:315–336
137. Rowlett VW, Margolin W (2013) The bacterial min system. Curr Biol 23:R553–R556
138. Kohyama S, Merino-Salomon A, Schwille P (2022) In vitro assembly, positioning and contraction of a division ring in minimal cells. Nat Commun 13:6098
139. Schwarz-Schilling M, Aufinger L, Muckl A, Simmel FC (2016) Chemical communication between bacteria and cell-free gene expression systems within linear chains of emulsion droplets. Integr Biol (Camb) 8:564–570
140. Lentini R, Santero SP, Chizzolini F, Cecchi D, Fontana J, Marchioretto M, Del Bianco C, Terrell JL, Spencer AC, Martini L, Forlin M, Assfalg M, Dalla Serra M, Bentley WE, Mansy

SS (2014) Integrating artificial with natural cells to translate chemical messages that direct E. coli behaviour. Nat Commun 5:4012
141. Lentini R, Martin NY, Forlin M, Belmonte L, Fontana J, Cornella M, Martini L, Tamburini S, Bentley WE, Jousson O, Mansy SS (2017) Two-way chemical communication between artificial and natural cells. ACS Cent Sci 3:117–123
142. Tang TD, Cecchi D, Fracasso G, Accardi D, Coutable-Pennarun A, Mansy SS, Perriman AW, Anderson JLR, Mann S (2018) Gene-mediated chemical communication in synthetic protocell communities. ACS Synth Biol 7:339–346
143. Halleran AD, Murray RM (2018) Cell-free and in vivo characterization of Lux, Las, and Rpa Quorum activation systems in E. coli. ACS Synth Biol 7:752–755
144. Adamala KP, Martin-Alarcon DA, Guthrie-Honea KR, Boyden ES (2017) Engineering genetic circuit interactions within and between synthetic minimal cells. Nat Chem 9:431–439
145. Din MO, Danino T, Prindle A, Skalak M, Selimkhanov J, Allen K, Julio E, Atolia E, Tsimring LS, Bhatia SN, Hasty J (2016) Synchronized cycles of bacterial lysis for in vivo delivery. Nature 536:81–85
146. Niederholtmeyer H, Stepanova V, Maerkl SJ (2013) Implementation of cell-free biological networks at steady state. Proc Natl Acad Sci U S A 110:15985–15990
147. Karzbrun E, Tayar AM, Noireaux V, Bar-Ziv RH (2014) Synthetic biology. Programmable on-chip DNA compartments as artificial cells. Science 345:829–832
148. Bartelt SM, Steinkuhler J, Dimova R, Wegner SV (2018) Light-guided motility of a minimal synthetic cell. Nano Lett 18:7268–7274
149. Carlier MF (1990) Actin polymerization and ATP hydrolysis. Adv Biophys 26:51–73
150. Carlier MF, Valentin-Ranc C, Combeau C, Fievez S, Pantoloni D (1994) Actin polymerization: regulation by divalent metal ion and nucleotide binding, ATP hydrolysis and binding of myosin. Adv Exp Med Biol 358:71–81
151. Carlier MF (1991) Nucleotide hydrolysis in cytoskeletal assembly. Curr Opin Cell Biol 3:12–17
152. Kato Y, Miyakawa T, Tanokura M (2018) Overview of the mechanism of cytoskeletal motors based on structure. Biophys Rev 10:571–581
153. Sunchu B, Cabernard C (2020) Principles and mechanisms of asymmetric cell division. Development 147:dev167650
154. Siton-Mendelson O, Bernheim-Groswasser A (2016) Toward the reconstitution of synthetic cell motility. Cell Adh Migr 10:461–474
155. Van Nies P, Westerlaken I, Blanken D, Salas M, Mencia M, Danelon C (2018) Self-replication of DNA by its encoded proteins in liposome-based synthetic cells. Nat Commun 9:1583
156. Libicher K, Hornberger R, Heymann M, Mutschler H (2020) In vitro self-replication and multicistronic expression of large synthetic genomes. Nat Commun 11:904
157. Miller OJ, Bernath K, Agresti JJ, Amitai G, Kelly BT, Mastrobattista E, Taly V, Magdassi S, Tawfik DS, Griffiths AD (2006) Directed evolution by in vitro compartmentalization. Nat Methods 3:561–570
158. Dodevski I, Markou GC, Sarkar CA (2015) Conceptual and methodological advances in cell-free directed evolution. Curr Opin Struct Biol 33:1–7
159. Hestericova M, Heinisch T, Alonso-Cotchico L, Marechal JD, Vidossich P, Ward TR (2018) Directed evolution of an artificial imine reductase. Angew Chem Int Ed Engl 57:1863–1868
160. Gao W, Bu N, Lu Y (2019) Efficient incorporation of unnatural amino acids into proteins with a robust cell-free system. Methods Protoc 2:16
161. Ranji Charna A, Des Soye BJ, Ntai I, Kelleher NL, Jewett MC (2022) An efficient cell-free protein synthesis platform for producing proteins with pyrrolysine-based noncanonical amino acids. Biotechnol J 17:e2200096
162. Martin RW, Des Soye BJ, Kwon YC, Kay J, Davis RG, Thomas PM, Majewska NI, Chen CX, Marcum RD, Weiss MG, Stoddart AE, Amiram M, Ranji Charna AK, Patel JR, Isaacs FJ, Kelleher NL, Hong SH, Jewett MC (2018) Cell-free protein synthesis from genomically

recoded bacteria enables multisite incorporation of noncanonical amino acids. Nat Commun 9: 1203
163. Dudley QM, Anderson KC, Jewett MC (2016) Cell-free mixing of Escherichia coli crude extracts to prototype and rationally engineer high-titer mevalonate synthesis. ACS Synth Biol 5:1578–1588
164. Karim AS, Jewett MC (2016) A cell-free framework for rapid biosynthetic pathway prototyping and enzyme discovery. Metab Eng 36:116–126
165. Voyvodic PL, Bonnet J (2020) Cell-free biosensors for biomedical applications. Curr Opin Biomed Eng 13:9–15
166. Pardee K, Green AA, Ferrante T, Cameron DE, Daleykeyser A, Yin P, Collins JJ (2014) Paper-based synthetic gene networks. Cell 159:940–954
167. Jung JK, Alam KK, Verosloff MS, Capdevila DA, Desmau M, Clauer PR, Lee JW, Nguyen PQ, Pasten PA, Matiasek SJ, Gaillard JF, Giedroc DP, Collins JJ, Lucks JB (2020) Cell-free biosensors for rapid detection of water contaminants. Nat Biotechnol 38:1451–1459

Adv Biochem Eng Biotechnol (2023) 185: 21–46
https://doi.org/10.1007/10_2023_226

Published online: 13 June 2023

Solid-Phase Cell-Free Protein Synthesis and Its Applications in Biotechnology

Mercedes Sánchez-Costa and Fernando López-Gallego

Contents

Abstract Cell-free systems for the in vitro production of proteins have revolutionized the synthetic biology field. In the last decade, this technology is gaining momentum in molecular biology, biotechnology, biomedicine and even education. Materials science has burst into the field of in vitro protein synthesis to empower the value of existing tools and expand its applications. In this sense, the combination of solid materials (normally functionalized with different biomacromolecules) together with cell-free components has made this technology more versatile and robust. In this chapter, we discuss the combination of solid materials with DNA and transcription–translation machinery to synthesize proteins within compartments, to

M. Sánchez-Costa (✉) and F. López-Gallego (✉)
Heterogeneous Biocatalysis Laboratory, Center for Cooperative Research in Biomaterials (CIC biomaGUNE), Basque Research and Technology Alliance (BRTA), Donostia-San Sebastián, Spain
e-mail: flopez@cicbiomagune.es

immobilize and purify in situ the nascent protein, to transcribe and transduce DNAs immobilized on solid surfaces, and the combination of all or some of these strategies.

Graphical Abstract

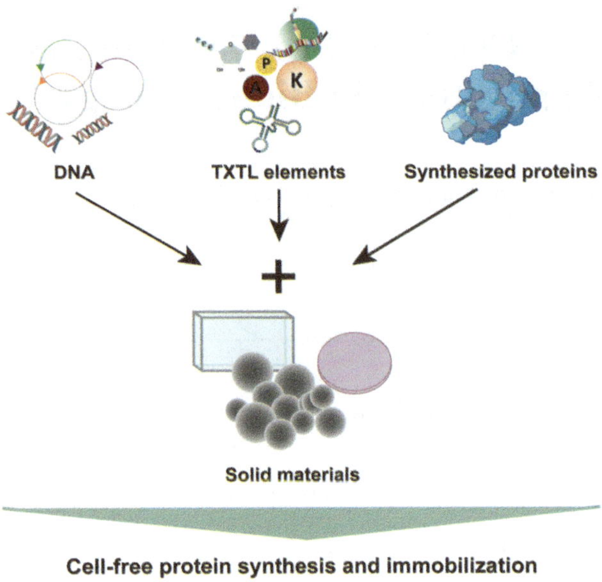

Keywords Biosensor, Cell-free protein synthesis, Compartmentalization, Hydrogels, Immobilization, Synthetic biology

1 Introduction

1.1 Overview

It has been six decades since Nirenberg and Matthaei [1] used *Escherichia coli* to isolate the transcription and translation machinery to in vitro synthesize proteins. This achievement revolutionized the biology and synthetic biology fields. In this pioneering work, they found a relationship between RNA and protein synthesis, expanding their knowledge of the translation process. Since then, it has been accomplished numerous advances in cell-free protein synthesis (CFPS), however, some bottlenecks remain unsolved, which encourages scientists and technologists to carry out further improvements. For example, the low yields of functional proteins and the lack of reproducibility hamper the implementation of CFPS beyond its investigation as proof of concept at the laboratory scale (Fig. 1).

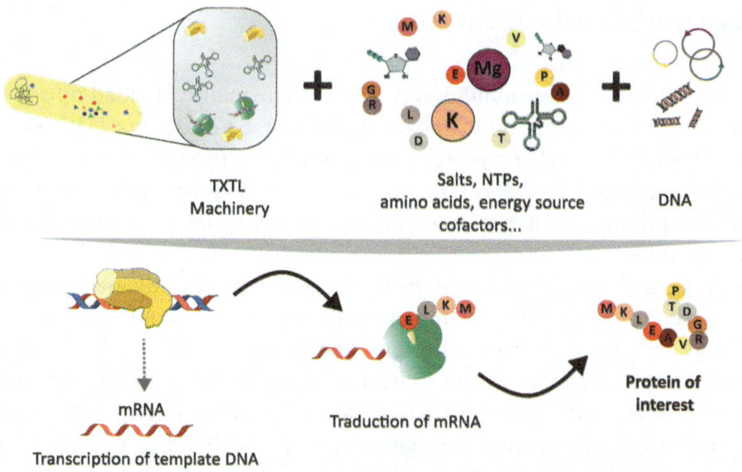

Fig. 1 Scheme of in vitro transcription and translation (TXTL) systems. The reaction mix contains the machinery for the cell-free protein synthesis obtained from the corresponding organism, including ribosomes, RNA polymerase and other enzymes, other exogenous supplements such as salts, NTPs, amino acids, etc., and the DNA template encoding the protein of interest. The reaction starts with the transcription of DNA to mRNA catalysed by the RNA polymerase, and the nascent mRNA is used as the template in the ribosome-mediated transduction to produce the final protein

Despite this fact, CFPS is a valid solution for the lab-scale production of toxic proteins where the viability of living organisms limits their expression or membrane proteins that usually are insoluble, being hard to purify [2]. This approach has underpinned some fundamental studies in biology to understand the functionality of protein complexes which are extremely difficult to produce in vivo [3]. Additionally, CFPS are open environments that permit the modification of the reaction components without affecting the genomics and metabolism of the organisms, enabling, for instance, the introduction of non-natural amino acids into proteins or their artificial post-translation modifications. Finally, CFPS can synthesize nascent proteins more rapidly than using living organisms, boosting prototyping studies to engineering enzymes, vaccines or even biosynthetic pathways.

Over the years, diverse factors have been studied and optimized [4, 5], including the methodology for obtaining cellular extracts [6, 7] and the generation of improved genetically modified strains [8, 9], as well as the successive synthesis reaction, from composition elements, their concentration and preparation and storage [10–13]. Moreover, DNA template sequences [14] have also been engineered and regeneration energy systems [15–18] have improved to increase the longevity of the process.

1.2 Types of Cell-Free Systems

Systems for cell-free protein synthesis (CFPS) can be mainly divided into two types; lysed-based extracts, commonly known as S30 or S12 extracts [19], or purified systems (PURE) [20], where protein synthesis is performed by pure His-tagged recombinant elements that form the transcription and translation machinery of *E. coli*. Both options present pros and cons, so the final goal will dictate the selection of one or the other.

While systems based on cell extracts (or lysates) have the advantage that they are simpler to prepare (obtaining a complete cell extract) and more versatile (cell-free extracts can be prepared from different cell lines), PURE systems require the expression and purification of the main transduction elements (initiation factors, elongation, release and recycling of ribosomes, T7 RNA polymerase, aminoacyl, tRNA synthetases and enzymes for energy regeneration) which is a very laborious task, but on the other hand,d they offer more control over the synthesis reaction environment.

1.2.1 Cellular Lines

Over the years, a variety of organisms have been employed in protein production with the cell-free methodology, being remarkable the *E. coli* bacteria [7, 21–24], likely due to easy handling, high protein yield of lysate and its good activity levels. However, its use shows the constraints to incorporate post-transcriptional modification in eukaryotic proteins. Among other used extracts, those from wheat germ (plants) [25–29] and rabbit reticulocyte (animals) [30–32], but also from mammalian cells (humans) [33] have been exploited to in vitro synthesize more complex proteins. Moreover, new organisms have been employed, including extreme prokaryotes such as *Thermus thermophilus* [34, 35].

Although most examples for CFPS have been performed in bulk, the unification of cell-free and materials science has recently boosted the emergence of new and improved tools for the in vitro synthesis of proteins in solid-phase, harnessing the features of the materials to create new functionalities and applications that go beyond the traditional CFPS in solution. In this chapter, we will attempt to summarize and exemplify how the main components of these reactions (DNA, cellular extracts, and protein products) have been combined with a plethora of materials to open new windows in the discovery and development of innovative applications.

2 Transcription–Translation Elements Compartmentalization

2.1 Entrapment

In living organisms, the machinery of transcription and translation (TX-TL) works in a high molecular density environment, which accelerates the biochemical processes. To maximize the potential of cell-free reactions, it is important to mimic these working conditions, which are normally achieved with small molecules added as crowding agents in synthesis reactions. The most employed crowding molecules are polymers such as PEG or Ficoll [36, 37]. Nonetheless, researchers have been lastly inspired by the nature to create artificial chassis that compartmentalize CFPS.

Biomaterials are gaining importance in numerous research fields [38]. These materials are conceived to interact with biological systems, thus they must be biocompatible and functional in the biological context (i.e. physiological conditions). Among the different types of biomaterials, hydrogels dominate the landscape of the applications. Hydrogels can be natural or semi-synthetic based on their nature, and they are characterized by having a crosslinked macromolecular network. They have a high swollen capacity that makes them able to absorb and release water and other fluids easily. Besides the good results in other applications like drug delivery, hydrogels are also excellent chassis for the entrapment of transcription and translation elements.

Reports exposing the use of hydrogels for protein synthesis can be widely found in the last decade. However, the exploration for alternative hydrogels to augment their functionality and decrease their negative impact caused by the polymer environment in the synthesis reactions is attracting several investigations. The use of a variety of matrixes as chassis for CFPS has been broadly investigated by Collete J. Whitefield et al. [39, 40]. Their studies include different polymers of different nature characterized by diverse polymerization chemistries mechanisms. This broad research encompasses polysaccharide-based (agarose, agar, xanthan, gum, hyaluronic acid, Gelzan™, and alginate) or amino acid-based gels (collagen, gelatin, and HydroMatrix™), also poloxamers (Pluronic acid F108 and F127), and finally the well-known polyacrylamide. The crosslink strength in these materials varies according to the type of interaction that will result in the final gel, hydrogen bonds, Van der Waals forces, ionic interactions driven by divalent metals, micelle aggregates, or covalent bonds. This property directly affects the final yield of the protein production, which makes that the same hydrogel with different crosslinking degrees can be either totally compatible with or deleterious for CFPS. To highlight some of the results reported by Collete J. Whitefield et al., up to 400% fluorescence signal is obtained when CFPS of m-Cherry was performed within agarose, agar, xanthan, and Gelzan™ in comparison with bulk reactions.

Interestingly, the crosslinked nature is not the only parameter affecting the in vitro protein synthesis since the same matrix provides dissimilar compatibility levels when the material preparation is performed in one or more steps. To make the

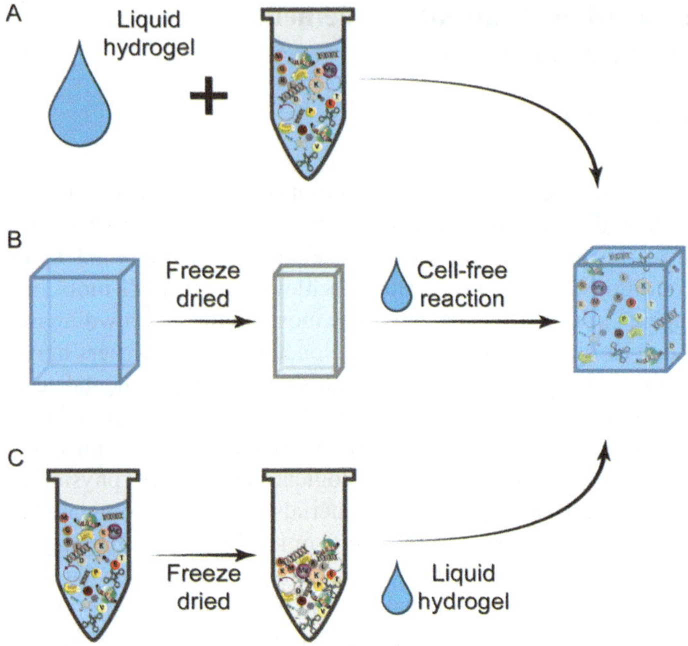

Fig. 2 Mechanisms for preparing hydrogels with entrapped CFPS. (**a**) Solutions of hydrogel and CFPS reaction are mixed before polymerization into the final hydrogel. (**b**) The polymerized hydrogel is freeze dried and rehydrated with the cell-free reaction mixture to form the final hydrogel. (**c**) The cell-free reaction is freeze dried and rehydrated with the liquid polymer to form the final hydrogel

hydrogel more compatible with the CFPS, the mixing process can be tuned as shown by Whitefield. In the study, the polymerization is evaluated from a premixed solution (polymer and cell-free components in one step) or from the lyophilized preparation of either the polymer or the cell-free synthesis reaction followed by the hydration with the non-lyophilized component that is left (two step) (Fig. 2). Two steps processes can prevent incompatibility requirements, as is the case for agar and agarose hydrogels that need high temperatures to be liquid, which can inactivate or degrade some cell-free components. In order to preserve the activity of CFPS, the working temperature range is too narrow, making the handling of the reactants [41] extremely difficult. Unfortunately, there is no universal protocol to prepare these matrixes, but this expands the possibilities for unsuited combinations.

Hydrogels are porous materials that allow for the movement of molecules within their framework. Molecular interactions within the hydrogel are influenced by the type of porosity as biomacromolecules can be too crowded to interact each other in small pores. Hence, the control of the polymer pore is a good tool to determine the best conditions to enhance the activity of all the components. On the other hand, the homogeneity presented by the hydrogel also impacts the yield of protein production, because clusters of entrapment materials can be formed preferentially in some areas.

Along with the effect on the reaction, the size of the pores can be used as a tool for controlling which compounds can internally or externally diffuse [42]. This diffusion mechanism also affects the nascent protein, which can move from the hydrogel to the bulk environment and vice versa. The ability to control the diffusion of the protein is a powerful tool in drug delivery applications. Hydrogels can be implanted in specific areas where the drug must be delivered. In this way, once the protein is produced, it can diffuse more precisely to the target region. This approach is especially interesting in biomedicine in which it is required to reduce the potential cross-reactions of the therapeutics.

Finally, the hydrogels have not to be just designed for supporting the CFPS elements, as polymerizing enzymes can be produced in vitro to trigger the polymerization of monomers present at the reaction mixture, forming a gel simultaneously with the action of the CFPS [43]. In this manner, it is possible to control the hydrogel generation and study the polymerization mechanisms in situ.

2.1.1 Cell-Free Protein Synthesis-Based Biosensors

One of the most paradigmatic applications of entrapped CFPS systems is their use as biosensors. This technology has boosted this field during the last decade. These have emerged as an alternative to the common microorganism-based biosensors that, although highly used, are sometimes poorly stable and their use is considered more hazardous due to the use of living organisms. Cell-free based biosensors can be employed for the finding of chemical compounds and organisms. Thus, their application is highly valuable in the environmental field, where the detection of pollutants is nowadays critical for humans and the fauna and flora, as well as the detection of pathogens for public health. Basically, these biosensors are designed to synthesize a reporter protein whose production is controlled by the presence or absence of a specific regulator, as they can be small molecules, other proteins or nucleic acid molecules, among others.

The genetic circuits to control the expression of the reporter proteins have evolved from basic designs to sophisticated strategies where to stand out the use of toehold riboswitches. This riboregulation of the gene expression is triggered by specific RNA molecules, whose absence keeps the ribosome binding site (RBS) blocked by a secondary structure formation that impedes the ribozyme attachment, leading to an OFF state. On the contrary, the triggering RNA molecule attachment induces a conformational change, allowing ribosomes to bind the RBS and the consequent ON state. To deepen into the riboswitch types in cell-free systems, the reader can attend to more specific reviews [44, 45].

Biosensors have to be economical devices that, ideally, may be used on-site avoiding the need for transporting samples to specialized laboratories. Therefore, keeping the cold chain is a common issue to face, especially when access to remote areas is required. As a potential solution, many researchers have proved that to freeze dry [46–48] the cellular lysates increase their stability at room temperature, expanding their durability for up to months. Thus, biosensors composed of a solid matrix where cell extracts are embedded and lyophilized altogether, are a promising tool for in situ detection of pollutants and for diagnostics.

Paper is one of the most assessed supports due to its low cost and easy handling. Numerous studies have focused on the development of paper-based biosensors based on CFPS able to detect hazardous metals such as Hg^{2+} or Pb^{2+} [49], drugs such as γ-hydroxybutyrate [50], chemicals such as phloroglucinol [51], and pathogens such as G II.4 Sydney virus genotype [52] and other norovirus [53], Sars-Cov-2 [54], syncytial virus classes A and B [55], or Zika [46, 56] among other demonstrations [57–60].

The majority of paper-based biosensors employed up to date are cellulose based, being remarkable the cellulose acetate matrix. However, its high demand related to their wide use in diagnostics devices and the likely binding of the components of cell extracts to the substrate fibres have augmented the search for alternative paper-based materials compatible with cell-free systems. An extensive screening included cellulose nitrate, nylon, Ashless cellulose filter paper, printer paper, and 100% cotton cellulose chromatography paper [54].

Although many advances have been reached within the field of cell-free paper-based biosensors, there is still room for improvement and for so, researchers are making efforts to discover optimized biosensors, as seeking new cheap and highly available paper derivates with great biocompatibility with cell-free components. Regarding the detection methodology, where fluorescence, colorimetric couple reaction and bioluminescence are the most used outputs signals, there are also concerning topics. In order to generalize the use of these devices in remote regions and whenever possible by non-specialized personnel, the readout signals have to be easy to interpret, however, some biosensors still need laboratory equipment to obtain the output values, especially when they are fluorescence-based. Finally, other parameters are of interest to keep investigations open such as the decrease of sensitivity thresholds and cross-reactivity, the development of more complex and specific genetic circuits to increase the precision of diagnostics procedures, and the multiplying of samples to rehydrate the CFPS reaction without the necessity for previous sample treatment.

2.2 *Immobilization*

CFPS reactions besides being entrapped, can also be immobilized on solid materials. Unlike cellular lysates, the transcription/translation (TX-TL) machinery expressed for PURE systems [20] that contain all its enzymes fused to histidine tags can be immobilized on surfaces activated with metal chelates (i.e. Ni^{2+}-NTA). Xiaoyu Zhou et al. [42] demonstrate that PURE elements are active when they are immobilized on polyacrylamide functionalized with Ni^{2+}-NTA. Other non-tagged components such as the ribosomes remain retained in the matrix thanks to the pore size of the polymer, as mentioned above, while the small molecules (i.e. amino acids and nucleotides) freely diffuse through the materials. This "synthetic cell" expands the life of PURE enzymes, resulting in a system active for 11 days when the replacement of small molecules is kept and up to 30 days in an updated version.

Investigations carried out with the immobilization of these TX-TL pure elements on nanoparticles (NPs) demonstrate higher output values for synthesized fluorescence proteins that free CFPS systems [60, 61]. As previously described by Thakur and co-workers [62], when enzymes belonging to an enzymatic cascade are immobilized on NPs, aggregates are generated, and a series of motives related to this aggregation may justify the higher activity of these enzymes such as the surface effect or a potential substrate channelling. These hypotheses are focused on higher performances of immobilized enzymes thanks to a better flux of substrate and enhanced kinetics. Nevertheless, the analysis focused on how much protein is actually produced revealed a non-increase yield in the process of synthesis. Although the immobilization of TX-TL elements negliglibyl increase the protein yield, it enhances the functional-unfunctional ratio of the nascent protein. Therefore, through the immobilization of pure elements, it is possible to overcome one of the bottlenecks in the implementation of cell-free synthesis as it is the low functional produced proteins in many cases while at the same time costs per experiment can be reduced since less pure elements are required for equivalent values of synthesized functional proteins.

Nonetheless, the main application explored with the use of beads functionalized with "nitrotriacetic acid (NTA)" coordinated to divalent metals is the anchoring of the nascent protein for its in situ purification, a topic that will be extended below.

Apart from the properties of the environment found in livings organisms, the perfect orchestration of processes that takes place in cells has also boosted investigations seeking for mimicking this organization. For example, compartmentalization of biosynthetic pathways is quite common in nature. The TX-TL process reviewed in this manuscript illustrates quite well the importance of such compartmentalization to enhance the performance of isolated biological machineries. While in the open bulk reactions normally run in laboratories both transcription and translation occur in the same solution, cells have them physically segregated. The transcription happens in the nucleus from where the mRNA migrates to be translated into a protein in the cytoplasm. The inherent semi-permeability of some hydrogels mimics such compartmentalization as recently proposed by Li and collages. In the proposed strategy termed ConCEiV (configurable compartmentation strategy for engineered in vitro molecular and cellular systems), the authors drive the compartmentalization designing hydrogels with a core of Polyacrylamide-N,N′-bis(acryloyl)cystamine surrounded by a thin agarose shell covalently attached to the core. This complex architecture was achieved in microfluidic devices. The core gel can be easily converted into a liquid phase by applying mild conditions, while the agarose shell remains jellified to control the pore size for the controlled diffusion of separated components (Fig. 3). This approach brings in vitro cell-free systems closer to cellular events. Moreover, the incorporation of magnetic nanoparticles into the agarose shell offers the capability of recovering and recycling the ConCEiV by magnetic separation and refreshing the external solution. This possibility expands the life cycles of the system up to five times, although the protein yield was decreased. Thus, one of

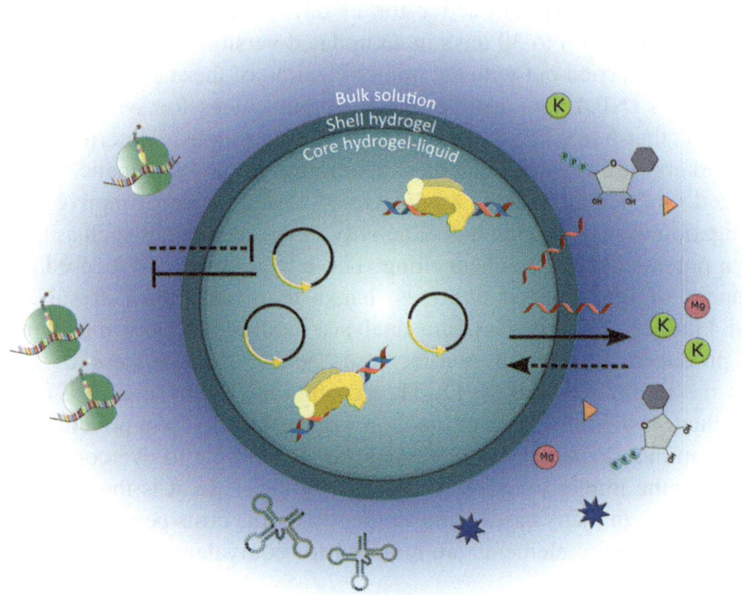

Fig. 3 Spatially compartmentalized transcription and translation reactions performed in hydrogels. In the core hydrogel, DNA and RNA polymerase are encapsulated. The rest of the components of a CFPS reaction are in the bulk solution. The exchange of elements between phases depends on the hydrogel porosity and the size of the elements forming the CFPS. The DNA and RNA polymerase are retained within the hydrogel while the ribosomes remain in the bulk solution. Once the mRNA is transcribed in the inner layer, it diffuses to the bulk solution to be used in the translation reaction. Arrows indicate components that can circulate between the layers and the T-shape arrows the ones that cannot

the main disadvantages of CFPS, which is the run out of components, is partially overcome [63].

This work proofs that the compartmentalization of TX-TL reactions into diverse solid environments can increase the yield of the nascent proteins, allows the manufacturing of protein-based biomaterials for controlled drug delivery, and facilitates the translation of CFPS-based biosensors form the lab to non-laboratory environments to monitor analytes in place.

3 Recovery and Purification of the Nascent Proteins Synthesized by CFPS

As it has been described and discussed in the previous section, the creation of protein-based biomaterials is highly significant in many research and applied fields. The use of cell-free technology provides a longstanding tool to obtain well folded proteins in a straightforward manner. Unfortunately, the vast protein content of the cellular extracts and PURE systems could interfere with the final performance of our protein of interest. This fact is particularly revelant for biocataysis applications since the cell extract composition and density may either inactive or inhibit the nascent enzyme. For this purpose, a purification step is normally implemented upon the in vitro synthesis of any target protein for their further use. However, the in vitro synthesis and purification of the nascent proteins in various steps is a costly and time-demanding procedure. Thus, adding solid materials to the CFPS reaction mix either before or after the protein synthesis enables the one-pot protein synthesis and purification in a wither concurrent or sequential manner, respectively. The only indispensable requirement for the solid material is that it must be inert as its interaction with the transcription and translation machinery must be negligible to avoid negative impacts on the CFPS. Hence, the surface chemistry of the solid support that captures the nascent protein must be biorthogonal with the cell extract that contains the TX-TL machinery.

Recombinant DNA technologies allow the incorporation of biorthogonal protein tags, which are normally genetically fuse to either the N- or C-terminus of the target protein. More rarely, these tags can also be inserted in internal spots of the sequence [64, 65]. Many of these tags can be used for the purification of one specific protein in a complex mixture such as a cell extract, thus this technology can be easily expanded to the in situ purification of tagged proteins synthesized in vitro through CFPS systems. Hence, the possibilities of combining materials and CFPS for protein synthesis and purification/immobilization are increasing, opening new avenues for the design of advanced biomaterials. Some of the strategies to immobilize/purify the nascent protein in vitro using functional solid materials are discussed in the following subsections.

3.1 Affinity Tags

Immobilization metal affinity chromatography (IMAC) is one of the most extensive techniques utilized for protein purification. This method is based on the high affinity of the side chains of some specific amino acids such as histidine, cysteine, or tryptophan for different metals. Among them, histidine stands out as being the most used by far. The introduction of tags based on poly-His peptide either at the N- or -C terminal ends of the protein allows the selective immobilization of proteins on solid surfaces activated with metal chelates through metal-histidine coordination bonds. Depending on the metal, the affinity of the protein for the solid material varies. Ni^{2}, Co^{2+}, Zn^{2+}, and Cu^{2+} are the most widely used, although other divalent metals can be found.

Based on purification technologies developed for purifying proteins expressed in living organisms, researchers are generating new materials to adapt the purification toolbox to nascent proteins synthesized in vitro. Some examples show the successful application of NTA-Ni^{2+} or IDA-Co^{2+} activated porous microbeads for in situ protein recovery upon its synthesis. The great specificity of the metal–histidine interaction and the fast reaction mechanism also contribute to exploit metal chelates in combination with CFPS reactions.

Due to the fact that the metal can be lixiviated from the solid materials during the one-pot protein synthesis and immobilization, which can negatively affect the performance of the TX-TL machinery, scientists are seeking alternatives to make the synthesis and immobilization process compatible. For example, unusual protein–material interactions based on His-tags are being developed. In 2019, the group of Bo Zheng [66] showed that it is possible to replace the metals with His-catchers attached to the solid surface to capture the nascent His-tagged protein, keeping the high affinity of the interaction. As proof, they synthesized a polyacrylamide gel incorporating a grafted His-tag aptamer, which allows the immobilization of the proteins from the PURE system remaining them active for transcription and translation reactions. Besides this application, we envision that similar aptamers (or antibodies) can be exploited for the recovery of tagged proteins synthesized in vitro, especially for those cases where metals inhibit/inactivate the biological activity of the nascent proteins. In the case that the metal activated solid materials do not lixiviate the metal but interact with the TX-TL machinery inactivating it, new architectures are being created to compartmentalize the protein synthesis and immobilization into two operation units within the same microfluidic device [67]. In this configuration, nascent proteins are produced in the synthesis unit and then flowed to a purification unit where they are bound to cross-linked, NTA-Ni-activated agarose beads for immobilization. Authors expect to be able to reuse the system just by replacing the agarose beads. In any case, when one designs a system for one-pot experiments where protein synthesis is followed by its immobilization, it is advisable to optimize the type and concentration of the immobilization support to maximize the protein yields and immobilization effectiveness.

Although less applied for protein immobilization and recovery, other amino acid tags can also interact with solid materials via a non-metal chelate binding serving to immobilize the nascent proteins. Benitez-Mateos et al. have widely explored some alternatives to in situ synthesized and immobilized proteins with amino acid tags [68] through; (1) absorption on silica nanoparticles through Lys-tags through the ionic interaction of the primary amine group of Lys and the negative charge of the silica surface, (2) formation of thiol-gold bonds between thiol groups of cys-tag and gold nanorods, (3) covalent attachment between the same thiol groups of the Cys-tags and the epoxy groups displayed at the surface of porous methacrylate beads. As shown, the Cys-tag provides a variety of potential interactions to use together with cell-free systems for protein production and immobilization [69]. The selectivity of the immobilization depends on the type of interaction, thus we need to consider the thermodynamics, kinetics, and selectivity of interaction pair between the nascent protein and the support to guarantee that only the nascent protein is bound to the solid materials, which can be further separated from the CFPS reaction media and used directly as a functionalized biomaterial (Fig. 4).

A extensive research has been done in protein immobilization as it is an enabling technology to increase protein stability. Throughout the years, the effects of the

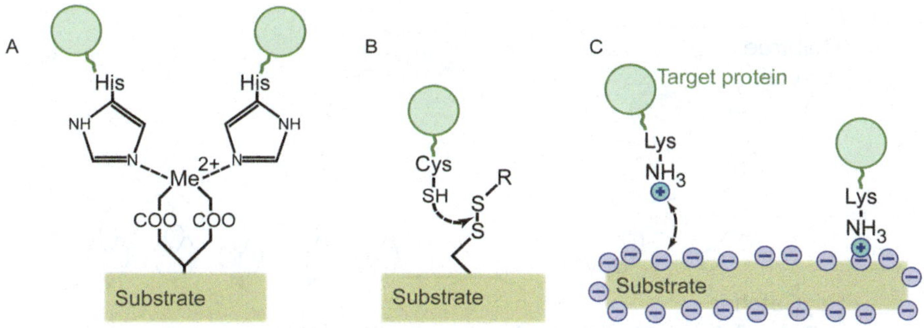

Fig. 4 Schematic representation of different amino acid tags and their interaction with the surface. (**a**) The surface is functionalized with divalent metals which interact with the histidine tag through coordination bonds. (**b**) The surface is activated with groups containing disulfide bonds that undergo a thiol-disulfide exchange with the cysteines of the tag. (**c**) The protein of interest interacts with the negative surface through ionic interactions with the positively charged lysines of the protein tag

types of immobilizations on enzyme activities have been investigated. Currently, we have access to a huge variety of mechanisms as well as solid materials to capture proteins. This linkage can occur through covalent and coordination bonds, ionic and hydrophobic interactions, Van der Waals forces, and specific tags and domains. Although currently many of these methodologies have been applied using in vivo produced proteins, this knowledge could be put into practice to expand the pathways for protein purification/immobilization from IVTT reactions. This approach would be also very interesting in the context of prototyping immobilized multi-enzyme systems to find the consensus immobilization chemistry for each of the enzymes involved in the biocatalytic cascade.

3.2 *Protein–Protein Tags*

In addition to peptide tags, protein domains can also act as binding motifs for the immobilization of nascent proteins upon their in vitro synthesis. While small peptides may be more efficient for protein purification/immobilization, the immobilization of proteins fused to large binding domains can be limited by steric hindrances between the binding domain and the reactive groups displayed at the surface of the solid supports.

In Nature, many organisms present proteins interacting with others by strong bonds, as well as spontaneous and stable isopeptide bonds. In the last years, this type of isopeptide connexion has inspired the search for proteins that can act as tags through the formation of these bonds. SpyTag/SpyCatcher and SnoopTag/SnoopCatcher dominate the research in isopeptide formation for protein–protein and protein–material interactions. This type of binding has been combined with the TX-TL methods for the selective immobilization of nascent proteins. The

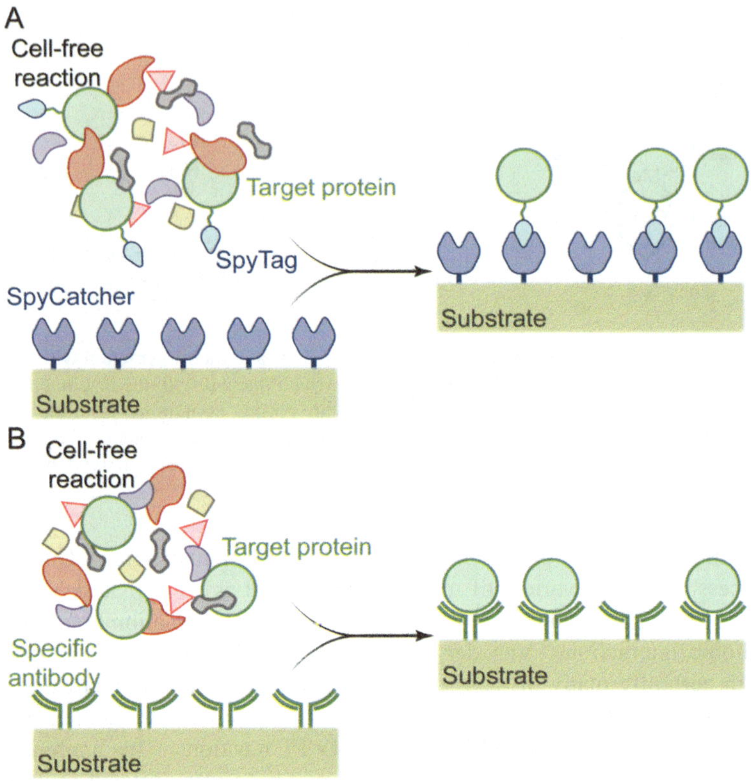

Fig. 5 Protein–protein and protein–epitope tags representations. The target protein in situ synthesized by CFPS reaction is captured by the surface functionalized with protein domains. (**a**) The surface is coated with the SpyCatcher protein that captures the nascent protein fused to the SpyTag. (**b**) Specific antibodies are spread over the surface to recognize the epitope of the nascent protein

SpyTag/SpyCatcher pair was developed from the immunoglobulin-like collagen adhesin module CnaB2 of the fibronectin-binding protein (Fbab) of *Streptococcus pyogenes* [70]. This module contains a stable isopeptide between one Lys and one Asp amino acid which greatly stabilizes the protein. The protein can be split into two components of 13 and 138 amino acids each, dubbed as Spy-Tag and Spy-Catcher, respectively. Remarkably, these two protein fragments can spontaneously form a peptide bond that irreversibly links them when they are incubated together. Similar to this pair, the one named SnoopTag/SnoopCatcher [71] is also derived from the division of the protein RrgA from *Streptococcus pneumoniae* into two modules that can bind via an isopeptide between one Lys and one Asn amino acid.

The introduction of the sequence encoding either the tag or the catcher into the DNA encoding the target protein, enables to in vitro synthesize tagged proteins that can be subsequently immobilized on surfaces functionalized with their cognate domain (either the catcher or the tag) (Fig. 5). Recently, Thornton, E L. et al. demonstrated the successful immobilization of nascent proteins through these two types of protein–protein interactions on solid materials. A hydrophobic glass surface was functionalized with BslA fused to the tag through protein self-assembly, as this protein assembles at hydrophobic/hydrophilic interfaces [72]. When CFPS systems produced a nascent protein fused to the catcher, this is immobilized on the glass surface decorated with the BslA fused to the tag through the formation of the corresponding isopeptide. This study demonstrates that functionalized materials with protein tags are selective since the target protein can be specifically captured from a protein mixture, with high affinity. Moreover, the strength of the interaction avoids the diffusion of proteins from the spots where they are anchored, a process that has been observed when immobilized proteins reversibly [73].

New tag/catcher pairs and variations have been developed in the last years [74, 75] and considering the proven orthogonality of some of them, new biomaterials can be created by activating the surface with different tags or catchers for trapping several enzymes at a time, which is a versatile tool that could permit the parallel synthesis and immobilization of enzymes without competition. In addition, we foresee new materials as the base for protein–protein scaffolds since there are still a few examples where several proteins are in situ synthesized and immobilized at the same time.

Still, there are plenty of other available but unexplored protein tags for one-pot CFPS and immobilization. Hence, we encourage the scientific community to bridge chemical biology and material sciences to exploit the potential of CFPS to decorate materials in a very precise manner.

3.3 *Epitope Tags*

Epitopes are the regions of molecules within the antigens that immune cells recognize. In the case of proteins or enzymes, specific amino acid sequences are recognized as epitopes by specific antibodies. Antigen–antibody interaction has been vastly exploited in numerous molecular biology techniques for protein detection and labelling such as western blot [76] or ELISA [77] (enzyme-linked immunosorbent assays). The epitope region can be intrinsic to the protein but can also be added as a fusion tag in the N- or C-terminus of proteins. Normally, the most antibody–antigen recognition is based on already existing antibodies like anti-His/His-tag or anti-FLAG/FLAG tag. For example, FLAG-tag has been used for the immobilization and characterization of the glycoside hydrolase GH78 from *Xylaria polymorpha* [78], using magnetic beads functionalized with Anti-FLAG for protein purification and quantification.

As it was mentioned above for other tags, antibody–epitope pair recognition can be employed to capture nascent proteins with the right epitope synthesized by CFPS systems (Fig. 5). The immobilization of antibodies in solid surfaces allows the creation of devices for further protein analysis (i.e. ligan binding). Some antibodies attached to different materials and combined with CFPS that can be found in the bibliography are the anti-GST (glutathione-S-transferase) [79], anti-Myc [80], and anti-HA (haemagglutinin) [81]. However, more and more advanced antibodies are commercially available including those fluorescence, bioluminescence, or enzyme tagged, so we expect that more applications will be developed in this modern field.

3.4 Protein Arrays

Microarrays are bidimensional chips based on solid surfaces where hundreds to thousands of biomolecules are spotted to run multiple analyses at a time [82]. From their discovery [83], they have revolutionized many research fields thanks to the vast number of results obtained in parallel in a short time. The most abundant microarrays found are DNA [84–89] and protein ones [90–93]. Proteins microarrays can be designed to run different types of analyses, such as functional analysis which can be biochemistry reaction studies, protein–protein, protein–DNA, and protein–RNA interactions assays, or analytical analysis for profiling the level expression of proteins, among others.

To generate these chips is indispensable to produce big amounts of proteins and, in many cases, a battery of variants to run exhaustive analysis. To this purpose, CFPS methodology facilities this task and so far, its use together with microfabrication technologies has enormously contributed to the revolution of the molecular biology. The concept of protein synthesis and immobilization was exploited for the first time in the production of new microarrays. Next, we illustrate the contribution of CFPS to the microarray technology with some remarkable examples. More than a decade ago, pioneering work by Mingyue, H and Ramachandran, N demonstrated that nascent proteins synthesized in vitro can be easily displayed in a microarray with excellent spatial resolution. These approaches were dubbed as PISA (protein in situ arrays) [94] and NAPPA (nucleic acid programmable protein array) [79, 95] methods. PISA method demonstrated the in vitro synthesis of His-tagged proteins and active enzymes that were subsequently immobilized on Ni^{2+}-NTA activated plates and magnetic beads. Similarly, NAPPA method proved the immobilization of GST-tagged proteins onto a glass surface coated by GST antibodies. These methodologies started to evolve years later [96, 97] up to date.

An example of this progress is shown with the co-tagging of proteins with the GST and HALO-tags for protein–protein interaction studies [98]. CFPS allows to synthesize both, GST-tagged target proteins, which attach to the surface, and HALO-tagged query proteins which are detected by the addition of a putative HALO-tag ligands labelled with fluorophores for the determination of protein–protein interactions. In this manner, a massive screening of protein interactions is performed faster and cheaper than by conventional methodologies.

A cost-effective and easy-to-handle method named TIPoHS (Translation and Immobilization of Protein on Hydrophobic Substrate) used polyvinylidene fluoride (PVDF) membranes [80], is commonly used in other techniques such as immunoblotting. This is a less sophisticated approach to make a microarray, where the fusion of a hydrophobic fluorescence protein to the C-terminus of target proteins allows the in situ attachment of the synthesized protein just by deposition of the CFPS reaction over the membrane. Although this method differs slightly from the microarray technology, a high density of spots localizing different CFPS reactions can be placed in a small piece of membrane, and by a simple fluorescence readout one can determine which proteins are correctly synthesized and folded in vitro.

Finally, these cell-free based protein-microarrays can be combined with analytical techniques. For example, with the SAMDI-MS (self-assembled monolayers for matrix-assisted laser desorption mass spectroscopy) [99]. This technique allows to detect post-translation modifications in the nascent proteins that are sequentially immobilized on the matrix. This immobilization is possible because the His-tagged nascent proteins react with the matrix surface functionalized with metal chelates. Through this method, the glycosylation degree of a mutant library in which the glycosylation motive has been included in each position of the Immunity Protein 7 (Im7) has been determined, overcoming the challenges of the quantification of post-translational modifications of individual proteins by classical labelling methodologies. Although this workflow included more than one step, it is one of the first high-throughput platforms for glycomics, which implies a high time decrease in the analysis of hundreds of protein variants.

These demonstrations highlight the potential of combining CFPS systems to empower even more the protein array technology.

4 DNA Immobilization

The last building block of the in vitro synthesis reactions that has been proven to be beneficially immobilized and used for functionalizing materials is DNA. The DNA harbours the genetic information that encodes either the protein to express or the regulatory elements for controlling transcription and translation reactions. The DNA employed as the template in CFPS reactions can have two forms, circular and linear, of which plasmids are the most commonly used due to their stability. However, during the last years, the use of linear DNAs has increased despite their tendency to be degraded by nucleases. The reason why linear DNA is preferred despite its lability is because it can be easily amplified by PCR, and its production does not require complex and laborious cloning procedures. Some of the new strategies developed for stabilizing linear templates included the addition of lambda bacteriophage GamS protein [100], which inhibits the RecBCD complex after its attachment, the supplementation with χ-site sequences oligos [101] (hotspots of crossover) which also bind the RecBCD complex, the addition of DNA-protection biomolecules [102], and the generation of genetically modified strains deficient in exonucleases [103], among others. Obtaining a high-quality linear DNA is an arduous process and a critical step to achieve good protein synthesis yields. The anchoring of linear DNA into solid surfaces provides some benefits for CFPS systems. Once the DNA is immobilized, it can be reused for subsequent CFPS reactions which reduce the total performing time. Moreover, the attachment offers protection against nucleases, augmenting the stability of the DNA molecules. When the particle size is appropriate for compartmentalization, the attachment of monoclonal DNAs allows the linking of the phenotype and the genotype. In addition, capturing the DNA in a solid matrix permits the control of its local concentration while it decreases its diffusion rate.

DAPA system [104] (DNA array to protein array) is one of the initial systems to produce enzymes from immobilized DNA. Nevertheless, many other variations have been lastly designed to take advantage of this approach. The most exploited interaction for preparing functionalized DNA materials is the biotin–streptavidin or homologues pair. This protein–small molecule interaction is one of the strongest non-covalent bonds found in Nature. Streptavidin is a homotetrameric enzyme which rapidly and selectively binds biotin. Moreover, the generated bond is very stable even under extreme biochemical conditions [105]. Altogether, this interaction is very popular for many research applications and fields. The most common strategy followed in the CFPS field is the coating of the material surfaces with the streptavidin protein which bind the DNA coupled to biotin molecules. Now, this immobilized DNA is biologically active for the RNA polymerase to accomplish the transcription process. This approach has been successfully employed for the in vitro synthesis of proteins using as the template biotinylated DNA immobilized over a variety of streptavidin activated supports such as sepharose [106], polystyrene [107], beads [108], and magnetic microbeads [109], as well as over other streptavidin homologues such as avidin [108]. Beyond this, other alternative linkers have been investigated with good findings, especially in the hydrogels discipline. DNA can be integrated into clay hydrogels, through the ionic interaction between the DNA and the charged clay nanodiscs [110]. DNA modified with acrydite can also be crosslinked with the acrylate group of a polymeric material to form a final DNA hydrogel [111]. In addition, more complex functionalization also results in functional DNA hydrogels. For instance, starting from hyaluronic acid, after a series of chemical reactions the generation of dibenzocyclooctyne groups allows the interaction with azide-DNA [112, 113].

The immobilization of the DNA does not only provide benefits in terms of the protein synthesis reactions, but also favours the in vitro formation of protein complexes due to the crowding of the individual components. As it is known to occur in cells, the association of subunits belonging to a protein complex is achieved due to different regulation mechanisms. The asseembly of these protein complexes can be controlled at both genetic level by organizing genes in clusters or operons [114] and physical level by confining the protein subunits in confined and crowded environments [115]. To prove these hypotheses in vitro, it was recently reported how the distribution of coding DNAs attached to silicon surface compartments follows a variety of patterns that can regulate the self-assembly of protein complexes. In addition, this study shows that the capture of the nascent subunit in the surroundings of the immobilized DNA enhances the formation of protein complexes. In this work, as previously described, the pair streptavidin–biotin is employed for DNA binding while the surface is covered with anti-HA (anti-haemagglutinin) for trapping expressed HA-tagged proteins. This approach allowed controlling the quantities and location of the DNA templates as well as the size and geometries of compartments were evaluated to deciphering the assembly order and yield of protein complexes during the in vitro formation of the baseplate wedge protein from the bacteriophage T4 and the *E. coli* RNAP.

It is important to remark that the immobilization of DNA is complementary to the methods described in previous sections. Therefore, both DNA and TX-TL machinery can be co-immobilized to expand the application of CFPS in developing new materials and technologies.

5 Closing Remarks and Future Perspectives

In the last years, the fusion of cell-free systems and solid materials is gaining momentum in the industry, medicine, biochemistry, and other disciplines. The combination of cell-free synthetic biology and material sciences has allowed the discovery of new applications such as a broad range of economical biosensors for the detection of many small molecules and organisms at the point of care or new hydrogel formulations for drug delivery with increased accuracy. The merge of these two disciplines has also assisted the protein arrays field, thanks to the high number of variants that can be expressed and plotted in one-step process. Besides, the immobilization of the target proteins into supports for their isolation from the CFPS reactions has improved their application in subsequent steps, permitting the expansion of their use in catalytic reactions or analytical techniques, among others. The good results moved researchers one step further, and initial wearable textile-based biosensors have been created, including a jacket or a face mask [116].

Although many achievements have been already reached, there is still room for improvement since the merge of these two fields has not been widely exploited yet, as well as some bottlenecks are still slowing down its development. In this sense, we expect to see novel and compatible materials to integrate CFPS with unexplored applications in fields as diverse as photonics or regenerative medicine. Aware of the limitation of the current CFPS and their incompatibility with some materials, we foresee that biocompatibility and biorthogonality are the major challenges to face for further integration of CFPS into new materials. These materials must go beyond being the mere support of the TX-TL machinery, but they must play an active role in the application. For example, magnetic or photonic materials may regulate the gene expression of a target protein under exposure to a magnetic field or light, respectively. As many of the usages have been developed with reporter proteins, we guess other proteins, enzymes or even complete complexes need to be interrogated. For example, prototyping of enzyme cascades immobilized of surface can be very valuable in applied biocatalysis to design more robust multi-functional biocatalysts. Complementarily, it seems upcoming investigations will try to be focused on more user-friendly preparations of CFPS, so they can be used by non-specialized personnel.

We hope that this chapter illustrates how relevant the complementation of material sciences cell-free synthetic biology is to bring innovation into biotechnology. Here are some of the more recent advances developed in the field, although it is not possible to cover in detail all the work performed on this subject, we consider that these successful examples can encourage the scientific community to continue researching at the interface between materials and CFPS.

References

1. Nirenberg MW, Matthaei JH (1961) The dependence of cell-free protein synthesis in *E. coli* upon naturally occurring or synthetic polyribonucleotides. Proc Natl Acad Sci U S A 47:1588–1602
2. Khambhati K, Bhattacharjee G, Gohil N, Braddick D, Kulkarni V, Singh V (2019) Exploring the potential of cell-free protein synthesis for extending the abilities of biological systems. Front Bioeng Biotechnol 7:248
3. Kuruma Y, Suzuki T, Ono S, Yoshida M, Ueda T (2012) Functional analysis of membranous Fo-a subunit of F1Fo-ATP synthase by in vitro protein synthesis. Biochem J 442:631–638
4. Garenne D, Thompson S, Brisson A, Khakimzhan A, Noireaux V (2021) The all-*E. coli* TXTL toolbox 3.0: new capabilities of a cell-free synthetic biology platform. Synth Biol (Oxf) 6: ysab017
5. Silverman AD, Kelley-Loughnane N, Lucks JB, Jewett MC (2019) Deconstructing cell-free extract preparation for in vitro activation of transcriptional genetic circuitry. ACS Synth Biol 8:403–414
6. Kim J, Copeland CE, Padumane SR, Kwon YC (2019) A crude extract preparation and optimization from a genomically engineered *Escherichia coli* for the cell-free protein synthesis system: practical laboratory guideline. Methods Protoc 2
7. Kwon YC, Jewett MC (2015) High-throughput preparation methods of crude extract for robust cell-free protein synthesis. Sci Rep 5:8663
8. Michel-Reydellet N, Woodrow K, Swartz J (2005) Increasing PCR fragment stability and protein yields in a cell-free system with genetically modified *Escherichia coli* extracts. J Mol Microbiol Biotechnol 9:26–34
9. Sato W, Sharon J, Deich C, Gaut N, Cash B, Engelhart AE, Adamala KP (2022) Akaby-cell-free protein expression system for linear templates. PLoS One 17:e0266272
10. Yang J, Cui Y, Cao Z, Ma S, Lu Y (2021) Strategy exploration for developing robust lyophilized cell-free systems. Biotechnol Notes 2:44–50
11. Hunt JP, Yang SO, Wilding KM, Bundy BC (2017) The growing impact of lyophilized cell-free protein expression systems. Bioengineered 8:325–330
12. Wilding KM, Zhao EL, Earl CC, Bundy BC (2019) Thermostable lyoprotectant-enhanced cell-free protein synthesis for on-demand endotoxin-free therapeutic production. New Biotechnol 53:73–80
13. Gregorio NE, Kao WY, Williams LC, Hight CM, Patel P, Watts KR, Oza JP (2020) Unlocking applications of cell-free biotechnology through enhanced shelf life and productivity of *E. coli* extracts. ACS Synth Biol 9:766–778
14. Karig DK, Iyer S, Simpson ML, Doktycz MJ (2012) Expression optimization and synthetic gene networks in cell-free systems. Nucleic Acids Res 40:3763–3774
15. Kim T-W, Keum J-W, Oh I-S, Choi C-Y, Kim H-C, Kim D-M (2007) An economical and highly productive cell-free protein synthesis system utilizing fructose-1, 6-bisphosphate as an energy source. J Biotechnol 130:389–393

16. Caschera F, Noireaux V (2014) Synthesis of 2.3 mg/ml of protein with an all *Escherichia coli* cell-free transcription–translation system. Biochimie 99:162–168
17. Anderson MJ, Stark JC, Hodgman CE, Jewett MC (2015) Energizing eukaryotic cell-free protein synthesis with glucose metabolism. FEBS Lett 589:1723–1727
18. Wang P-H, Fujishima K, Berhanu S, Kuruma Y, Jia TZ, Khusnutdinova AN, Yakunin AF, McGlynn SE (2019) A bifunctional polyphosphate kinase driving the regeneration of nucleoside triphosphate and reconstituted cell-free protein synthesis. ACS Synth Biol 9:36–42
19. Kim TW, Keum JW, Oh IS, Choi CY, Park CG, Kim DM (2006) Simple procedures for the construction of a robust and cost-effective cell-free protein synthesis system. J Biotechnol 126: 554–561
20. Shimizu Y, Kanamori T, Ueda T (2005) Protein synthesis by pure translation systems. Methods 36:299–304
21. Mullin AC, Slouka T, Oza JP (2022) Simple extract preparation methods for *E. coli*-based cell-free expression. Methods Mol Biol:51–64
22. Lehr F-X, Hanst M, Vogel M, Kremer J, Göringer HU, Suess B, Koeppl H (2019) Cell-free prototyping of AND-logic gates based on heterogeneous RNA activators. ACS Synth Biol 8: 2163–2173
23. Owens EM, Gussin GN (1983) Differential binding of RNA polymerase to the pRM and pR promoters of bacteriophage lambda. Gene 23:157–166
24. Shin J, Noireaux V (2012) An E. coli cell-free expression toolbox: application to synthetic gene circuits and artificial cells. ACS Synth Biol 1:29–41
25. Takai K, Sawasaki T, Endo Y (2010) The wheat-germ cell-free expression system. Curr Pharm Biotechnol 11:272–278
26. Okimune K-i, Nagy SK, Hataya S, Endo Y, Takasuka TE (2020) Reconstitution of Drosophila and human chromatins by wheat germ cell-free co-expression system. BMC Biotechnol 20:1–12
27. Novikova IV, Sharma N, Moser T, Sontag R, Liu Y, Collazo MJ, Cascio D, Shokuhfar T, Hellmann H, Knoblauch M (2018) Protein structural biology using cell-free platform from wheat germ. Adv Struct Chem Imaging 4:1–13
28. Yamaoka Y, Jeremiah SS, Funabashi R, Miyakawa K, Morita T, Mihana Y, Kato H, Ryo A (2022) Characterization and utilization of disulfide-bonded SARS-CoV-2 receptor binding domain of spike protein synthesized by wheat germ cell-free production system. Viruses 14: 1461
29. Abe S, Tanaka J, Kojima M, Kanamaru S, Hirata K, Yamashita K, Kobayashi A, Ueno T (2022) Cell-free protein crystallization for nanocrystal structure determination. Sci Rep 12: 16031
30. Ludgate L, Liu K, Luckenbaugh L, Streck N, Eng S, Voitenleitner C, Delaney IV WE, Hu J (2016) Cell-free hepatitis B virus capsid assembly dependent on the core protein C-terminal domain and regulated by phosphorylation. J Virol 90:5830–5844
31. Nylander A, Leznicki P, Vidovic K, High S, Olsson ML (2020) SMIM1, carrier of the Vel blood group, is a tail-anchored transmembrane protein and readily forms homodimers in a cell-free system. Biosci Rep 40
32. Douthwaite JA (2012) Eukaryotic ribosome display selection using rabbit reticulocyte lysate. Methods Protoc:45–57
33. Majumder S, Hsu Y-Y, Moghimianavval H, Andreas M, Giessen TW, Luxton GG, Liu AP (2022) In vitro synthesis and reconstitution using mammalian cell-free lysates enables the systematic study of the regulation of LINC complex assembly. Biochemistry 61:1495–1507
34. Uzawa T, Yamagishi A, Oshima T (2003) Continuous cell-free protein synthesis directed by messenger DNA and catalyzed by extract of *Thermus thermophilus* HB27. Biosci Biotechnol Biochem 67:639–642
35. Zhou Y, Asahara H, Gaucher EA, Chong S (2012) Reconstitution of translation from Thermus thermophilus reveals a minimal set of components sufficient for protein synthesis at high

temperatures and functional conservation of modern and ancient translation components. Nucleic Acids Res 40:7932–7945
36. Ge X, Luo D, Xu J (2011) Cell-free protein expression under macromolecular crowding conditions. PLoS One 6:e28707
37. Vezeau GE, Salis HM (2021) Tuning cell-free composition controls the time delay, dynamics, and productivity of TX-TL expression. ACS Synth Biol 10:2508–2519
38. Xie B, Ding Y-F, Shui M, Yue L, Gao C, Wyman IW, Wang R (2022) Supramolecular biomaterials for bio-imaging and imaging-guided therapy. Eur J Nucl Med Mol Imaging 49: 1200–1210
39. Whitfield CJ, Banks AM, Dura G, Love J, Fieldsend JE, Goodchild SA, Fulton DA, Howard TP (2019) Cell-free genetic devices confer autonomic and adaptive properties to hydrogels. bioRxiv:2019.12.12.872622
40. Whitfield CJ, Banks AM, Dura G, Love J, Fieldsend JE, Goodchild SA, Fulton DA, Howard TP (2020) Cell-free protein synthesis in hydrogel materials. Chem Commun 56:7108–7111
41. Benítez-Mateos AI, Zeballos N, Comino N, Moreno de Redrojo L, Randelovic T, López-Gallego F (2020) Microcompartmentalized cell-free protein synthesis in hydrogel μ-channels. ACS Synth Biol 9:2971–2978
42. Zhou X, Wu H, Cui M, Lai SN, Zheng B (2018) Long-lived protein expression in hydrogel particles: towards artificial cells. Chem Sci 9:4275–4279
43. Huang A, Nguyen PQ, Stark JC, Takahashi MK, Donghia N, Ferrante T, Dy AJ, Hsu KJ, Dubner RS, Pardee K (2018) BioBits™ explorer: a modular synthetic biology education kit. Sci Adv 4:eaat5105
44. Tabuchi T, Yokobayashi Y (2021) Cell-free riboswitches. RSC Chem Biol 2:1430–1440
45. Chushak Y, Harbaugh S, Zimlich K, Alfred B, Chávez J, Kelley-Loughnane N (2021) Characterization of synthetic riboswitch in cell-free protein expression systems. RNA Biol 18:1727–1738
46. Pardee K, Green AA, Takahashi MK, Braff D, Lambert G, Lee JW, Ferrante T, Ma D, Donghia N, Fan M (2016) Rapid, low-cost detection of zika virus using programmable biomolecular components. Cell 165:1255–1266
47. Pardee K, Slomovic S, Nguyen PQ, Lee JW, Donghia N, Burrill D, Ferrante T, McSorley FR, Furuta Y, Vernet A (2016) Portable, on-demand biomolecular manufacturing. Cell 167:248–259. e12
48. Smith MT, Berkheimer SD, Werner CJ, Bundy BC (2014) Lyophilized Escherichia coli-based cell-free systems for robust, high-density, long-term storage. BioTechniques 56:186–193
49. Zhang Y, Zhao C, Bi H, Zhang X, Xue B, Li C, Wang S, Yang X, Qiu Z, Wang J (2022) A cell-free paper-based biosensor dependent on allosteric transcription factors (aTFs) for on-site detection of harmful metals Hg2+ and Pb2+ in water. J Hazard Mater 438:129499
50. Gräwe A, Dreyer A, Vornholt T, Barteczko U, Buchholz L, Drews G, Ho UL, Jackowski ME, Kracht M, Lüders J, Bleckwehl T, Rositzka L, Ruwe M, Wittchen M, Lutter P, Müller K, Kalinowski J (2019) A paper-based, cell-free biosensor system for the detection of heavy metals and date rape drugs. PLoS One 14:e0210940
51. Meyer A, Saaem I, Silverman A, Varaljay VA, Mickol R, Blum S, Tobias AV, Schwalm III ND, Mojadedi W, Onderko E (2019) Organism engineering for the bioproduction of the triaminotrinitrobenzene (TATB) precursor phloroglucinol (PG). ACS Synth Biol 8:2746–2755
52. Ma D, Shen L, Wu K, Diehnelt CW, Green AA (2018) Low-cost detection of norovirus using paper-based cell-free systems and synbody-based viral enrichment. Synth Biol 3:ysy018
53. Sun Q, Cao M, Zhang X, Wang M, Ma Y, Wang J (2021) A simple and low-cost paper-based colorimetric method for detecting and distinguishing the GII. 4 and GII. 17 genotypes of norovirus. Talanta 225:121978
54. Hunt JP, Zhao EL, Free TJ, Soltani M, Warr CA, Benedict AB, Takahashi MK, Griffitts JS, Pitt WG, Bundy BC (2022) Towards detection of SARS-CoV-2 RNA in human saliva: a

paper-based cell-free toehold switch biosensor with a visual bioluminescent output. New Biotechnol 66:53–60
55. Cao M, Sun Q, Zhang X, Ma Y, Wang J (2021) Detection and differentiation of respiratory syncytial virus subgroups A and B with colorimetric toehold switch sensors in a paper-based cell-free system. Biosens Bioelectron 182:113173
56. Karlikow M, da Silva SJR, Guo Y, Cicek S, Krokovsky L, Homme P, Xiong Y, Xu T, Calderón-Peláez M-A, Camacho-Ortega S (2022) Field validation of the performance of paper-based tests for the detection of the zika and chikungunya viruses in serum samples. Nat Biomed Eng 6:246–256
57. Duyen TTM, Matsuura H, Ujiie K, Muraoka M, Harada K, Hirata K (2017) Based colorimetric biosensor for antibiotics inhibiting bacterial protein synthesis. J Biosci Bioeng 123:96–100
58. Matsuura H, Ujiie K, Duyen TTM, Izutsu K, Fujiyama K, Maeda S, Harada K, Hirata K (2019) Development of a paper-based luminescence bioassay for therapeutic monitoring of aminoglycosides: a proof-of-concept study. Appl Biochem Biotechnol 189:798–809
59. Beabout K, Ehrenworth Breedon AM, Blum SM, Miklos AE, Lux MW, Chávez JL, Goodson MS (2022) Detection of bile acids in complex matrices using a transcription factor-based biosensor. ACS Biomater Sci Eng
60. Selim AS, Perry JM, Nasr MA, Pimprikar JM, Shih SC (2022) A synthetic biosensor for detecting putrescine in beef samples. ACS Applied Bio Materials 5:5487–5496
61. Thakur M, Breger JC, Susumu K, Oh E, Spangler JR, Medintz IL, Walper SA, Ellis GA (2022) Self-assembled nanoparticle-enzyme aggregates enhance functional protein production in pure transcription-translation systems. PLoS One 17:e0265274
62. Vranish JN, Ancona MG, Oh E, Susumu K, Medintz IL (2017) Enhancing coupled enzymatic activity by conjugating one enzyme to a nanoparticle. Nanoscale 9:5172–5187
63. Li L, Zhang R, Chen L, Tian X, Li T, Pu B, Ma C, Ba F, Ba F, Xiong C, Shi Y, Mi X, Li J, Keasling JD, Zhang J, Liu Y (2022) Permeability-engineered compartmentalization enables invitro reconstitution of sustained synthetic biology systems. Adv Sci 9:2203652
64. Zordan RE, Beliveau BJ, Trow JA, Craig NL, Cormack BP (2015) Avoiding the ends: internal epitope tagging of proteins using transposon Tn7. Genetics 200:47–58
65. Oesterle S, Roberts TM, Widmer LA, Mustafa H, Panke S, Billerbeck S (2017) Sequence-based prediction of permissive stretches for internal protein tagging and knockdown. BMC Biol 15:1–17
66. Lai SN, Zhou X, Ouyang X, Zhou H, Liang Y, Xia J, Zheng B (2019) Artificial cells capable of long-lived protein synthesis by using aptamer grafted polymer hydrogel. ACS Synth Biol 9: 76–83
67. Xiao X, Zhou Y, Sun Y, Wang Q, Liu J, Huang J, Zhu X, Yang X, Wang K (2018) Integration of cell-free protein synthesis and purification in one microfluidic chip for on-demand production of recombinant protein. Biomicrofluidics 12:054102
68. Benítez-Mateos AI, Llarena I, Sánchez-Iglesias A, López-Gallego F (2018) Expanding one-pot cell-free protein synthesis and immobilization for on-demand manufacturing of biomaterials. ACS Synth Biol 7:875–884
69. Kamiya N, Ohama Y, Minamihata K, Wakabayashi R, Goto M (2018) Liquid marbles as an easy-to-handle compartment for cell-free synthesis and in situ immobilization of recombinant proteins. Biotechnol J 13:1800085
70. Zakeri B, Fierer JO, Celik E, Chittock EC, Schwarz-Linek U, Moy VT, Howarth M (2012) Peptide tag forming a rapid covalent bond to a protein, through engineering a bacterial adhesin. Proc Natl Acad Sci U S A 109:E690–E6E7
71. Veggiani G, Nakamura T, Brenner MD, Gayet RV, Yan J, Robinson CV, Howarth M (2016) Programmable polyproteams built using twin peptide superglues. Proc Natl Acad Sci U S A 113:1202–1207
72. Hobley L, Ostrowski A, Rao FV, Bromley KM, Porter M, Prescott AR, MacPhee CE, Van Aalten DM, Stanley-Wall NR (2013) BslA is a self-assembling bacterial hydrophobin that coats the *Bacillus subtilis* biofilm. Proc Natl Acad Sci U S A 110:13600–13605

73. Diamanti E, Arana-Peña S, Ramos-Cabrer P, Comino N, Carballares D, Fernandez-Lafuente R, López-Gallego F (2022) Intraparticle macromolecular migration alters the structure and function of proteins reversibly immobilized on porous microbeads. Adv Mater Interfaces 9
74. Tan LL, Hoon SS, Wong FT (2016) Kinetic controlled tag-catcher interactions for directed covalent protein assembly. PLoS One 11:e0165074
75. Keeble AH, Yadav VK, Ferla MP, Bauer CC, Chuntharpursat-Bon E, Huang J, Bon RS, Howarth M (2022) DogCatcher allows loop-friendly protein-protein ligation. Cell Chem Biol 29:339–350. e10
76. Towbin H, Staehelin T, Gordon J (1979) Electrophoretic transfer of proteins from polyacrylamide gels to nitrocellulose sheets: procedure and some applications. Proc Natl Acad Sci U S A 76:4350–4354
77. Engvall E, Perlmann P (1971) Enzyme-linked immunosorbent assay (ELISA) quantitative assay of immunoglobulin G. Immunochemistry 8:871–874
78. Knauer JF, Liers C, Hahn S, Wuestenhagen DA, Zemella A, Kellner H, Haueis L, Hofrichter M, Kubick S (2022) Cell-free production of the bifunctional glycoside hydrolase GH78 from *Xylaria polymorpha*. Enzym Microb Technol 161:110110
79. Ramachandran N, Hainsworth E, Bhullar B, Eisenstein S, Rosen B, Lau AY, Walter JC, LaBaer J (2004) Self-assembling protein microarrays. Science 305:86–90
80. Bhide M, Natarajan S, Hresko S, Aguilar C, Bencurova E (2014) Rapid in vitro protein synthesis pipeline: a promising tool for cost-effective protein array design. Mol BioSyst 10: 1236–1245
81. Vonshak O, Divon Y, Förste S, Garenne D, Noireaux V, Lipowsky R, Rudorf S, Daube SS, Bar-Ziv RH (2020) Programming multi-protein assembly by gene-brush patterns and two-dimensional compartment geometry. Nat Nanotechnol 15:783 791
82. Hall DA, Ptacek J, Snyder M (2007) Protein microarray technology. Mech Ageing Dev 128: 161–167
83. Fodor SP, Read JL, Pirrung MC, Stryer L, Lu AT, Solas D (1991) Light-directed, spatially addressable parallel chemical synthesis. Science 251:767–773
84. Scionti F, Arbitrio M, Caracciolo D, Pensabene L, Tassone P, Tagliaferri P, Di Martino MT (2022) Integration of DNA microarray with clinical and genomic data. Methods Mol Biol:239–248
85. Wöhrle J, Krämer SD, Meyer PA, Rath C, Hügle M, Urban GA, Roth G (2020) Digital DNA microarray generation on glass substrates. Sci Rep 10:1–9
86. Schneeberg A, Ehricht R, Slickers P, Baier V, Neubauer H, Zimmermann S, Rabold D, Lübke-Becker A, Seyboldt C (2015) DNA microarray-based PCR ribotyping of *Clostridium difficile*. J Clin Microbiol 53:433–442
87. Ma X, Li Y, Liang Y, Liu Y, Yu L, Li C, Liu Q, Chen L (2020) Development of a DNA microarray assay for rapid detection of fifteen bacterial pathogens in pneumonia. BMC Microbiol 20:1–13
88. Bozdech Z, Mok S, Gupta AP (2013) DNA microarray-based genome-wide analyses of *Plasmodium* parasites. Methods Mol Biol:189–211
89. Huan H, Zhang K, He J, Zhang J (2021) A DNA microarray assay for authenticating five important marine mammal species in food and feed. Food Chem 348:129136
90. Barderas R, Srivastava S, LaBaer J (2021) Protein microarray-based proteomics for disease analysis. Methods Mol Biol:3–6
91. Husain B, Paduchuri S, Ramani SR, Martinez-Martin N (2019) Extracellular protein microarray technology for high throughput detection of low affinity receptor-ligand interactions. J Vis Exp
92. Kilb N, Herz T, Burger J, Woehrle J, Meyer PA, Roth G (2019) Protein microarray copying: easy on-demand protein microarray generation compatible with fluorescence and label-free real-time analysis. Chembiochem 20:1554–1562

93. Du P-X, Chou Y-Y, Santos HM, Keskin BB, Hsieh M-H, Ho T-S, Wang J-Y, Lin Y-L, Syu G-D (2021) Development and application of human coronavirus protein microarray for specificity analysis. Anal Chem 93:7690–7698
94. He M, Taussig MJ (2001) Single step generation of protein arrays from DNA by cell-free expression and in situ immobilisation (PISA method). Nucleic Acids Res 29:e73–e73
95. LaBaer J, Ramachandran N (2005) Protein microarrays as tools for functional proteomics. Curr Opin Chem Biol 9:14–19
96. Lee K-H, Kwon Y-C, Yoo SJ, Kim D-M (2010) Ribosomal synthesis and in situ isolation of peptide molecules in a cell-free translation system. Protein Expr Purif 71:16–20
97. Lee K-H, Lee K-Y, Byun J-Y, Kim B-G, Kim D-M (2012) On-bead expression of recombinant proteins in an agarose gel matrix coated on a glass slide. Lab Chip 12:1605–1610
98. Tang Y, Qiu J, Machner M, LaBaer J (2017) Discovering protein-protein interactions using nucleic acid programmable protein arrays. Curr Protoc Cell Biol 74:15.21.1–15.21.14
99. Techner J-M, Kightlinger W, Lin L, Hershewe J, Ramesh A, DeLisa MP, Jewett MC, Mrksich M (2019) High-throughput synthesis and analysis of intact glycoproteins using SAMDI-MS. Anal Chem 92:1963–1971
100. Garamella J, Marshall R, Rustad M, Noireaux V (2016) The all *E. coli* TX-TL toolbox 2.0: a platform for cell-free synthetic biology. ACS Synth Biol 5:344–355
101. Marshall R, Maxwell CS, Collins SP, Beisel CL, Noireaux V (2017) Short DNA containing χ sites enhances DNA stability and gene expression in *E. coli* cell-free transcription–translation systems. Biotechnol Bioeng 114:2137–2141
102. Yim SS, Johns NI, Noireaux V, Wang HH (2020) Protecting linear DNA templates in cell-free expression systems from diverse bacteria. ACS Synth Biol 9:2851–2855
103. Azad MS, Batista AC, Faulon J-L, Beisel CL, Bonnet J, Kushwaha M (2022) Cell-Free protein synthesis from exonuclease-deficient cellular extracts utilizing linear DNA templates. J Vis Exp:e64236
104. He M, Stoevesandt O, Palmer EA, Khan F, Ericsson O, Taussig MJ (2008) Printing protein arrays from DNA arrays. Nat Methods 5:175–177
105. Dundas CM, Demonte D, Park S (2013) Streptavidin–biotin technology: improvements and innovations in chemical and biological applications. Appl Microbiol Biotechnol 97:9343–9353
106. Tanaka Y, Shimizu Y (2015) Integration of a reconstituted cell-free protein-synthesis system on a glass microchip. Anal Sci 31:67–71
107. Sierra AMR, Arold ST, Grünberg R (2022) Efficient multi-gene expression in cell-free droplet microreactors. PLoS One 17:e0260420
108. Iizuka R, Tahara K, Matsueda A, Tsuda S, Yoon DH, Sekiguchi T, Shoji S, Funatsu T (2022) Selection of green fluorescent proteins by in vitro compartmentalization using microbead-display libraries. Biochem Eng J 187:108627
109. Lee K-Y, Lee K-H, Park J-W, Kim D-M (2012) Flexible programming of cell-free protein synthesis using magnetic bead-immobilized plasmids. PLoS One 7:e34429
110. Jiao Y, Liu Y, Luo D, Huck WT, Yang D (2018) Microfluidic-assisted fabrication of clay microgels for cell-free protein synthesis. ACS Appl Mater Interfaces 10:29308–29313
111. Cui J, Wu D, Sun Q, Yang X, Wang D, Zhuang M, Zhang Y, Gan M, Luo D (2020) A PEGDA/DNA hybrid hydrogel for cell-free protein synthesis. Front Chem 8:28
112. Köhler T, Heida T, Hoefgen S, Weigel N, Valiante V, Thiele J (2020) Cell-free protein synthesis and in situ immobilization of deGFP-MatB in polymer microgels for malonate-to-malonyl CoA conversion. RSC Adv 10:40588–40596

113. Heida T, Köhler T, Kaufmann A, Männel MJ, Thiele J (2020) Cell-free protein synthesis in bifunctional hyaluronan microgels: a strategy for in situ immobilization and purification of his-tagged proteins. ChemSystemsChem 2:e1900058
114. Wells JN, Bergendahl LT, Marsh JA (2016) Operon gene order is optimized for ordered protein complex assembly. Cell Rep 14:679–685
115. Homouz D, Perham M, Samiotakis A, Cheung MS, Wittung-Stafshede P (2008) Crowded, cell-like environment induces shape changes in aspherical protein. Proc Natl Acad Sci U S A 105:11754–11759
116. Nguyen PQ, Soenksen LR, Donghia NM, Angenent-Mari NM, de Puig H, Huang A, Lee R, Slomovic S, Galbersanini T, Lansberry G (2021) Wearable materials with embedded synthetic biology sensors for biomolecule detection. Nat Biotechnol 39:1366–1374

Adv Biochem Eng Biotechnol (2023) 185: 47–58
https://doi.org/10.1007/10_2023_233

Published online: 11 August 2023

Cell-Free Protein Synthesis of Metalloproteins

Jamin Koo

Contents

Abstract Metalloproteins, proteins containing metal atoms or clusters within their structures, are critical for various biological functions across all domains of life. More than hundreds of different types have been discovered, which conduct various roles such as transportation of O_2, catalyzing chemical reactions, sensing environmental changes, and relaying electrons. Metalloprotein molecules incorporate a variety of metal atoms, coordinated to specific amino acid residues that affect their conformation and functionality. The process of metal incorporation typically occurs during or post-protein folding, often requiring chaperones for metal ion delivery and quality control. Progress in understanding metal incorporation and metalloprotein functionality has been enhanced by cell-free protein synthesis (CFPS) methods that offer direct control over the synthesis environment. This chapter reviews the diverse applications of CFPS methods in metalloprotein research, encompassing structure–function studies, protein engineering, and creation of artificial metalloproteins. Examples demonstrating the utility and advances brought about by CFPS in

J. Koo (✉)
Department of Chemical Engineering, Hongik University, Seoul, Republic of Korea
e-mail: jaminkoo@hongik.ac.krf

synthetic biology, electrochemistry, and drug discovery are highlighted. Despite remarkable progress, challenges remain in optimizing and advancing the CFPS methods, underscoring the need for future explorations in this transformative approach to metalloprotein study and engineering.

Graphical Abstract

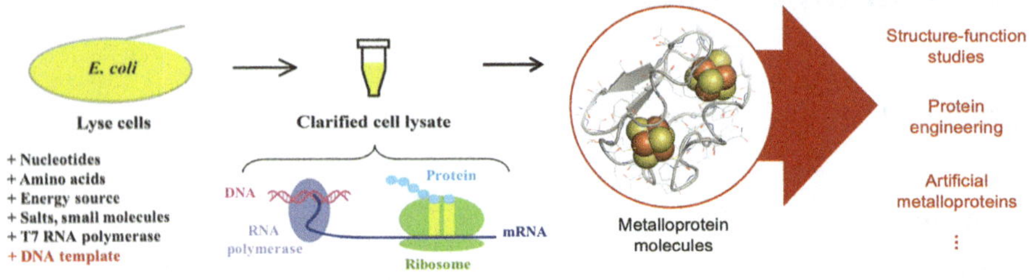

Keywords Cell-free, Chaperons, Maturation, Metalloproteins, Synthetic biology

1 Introduction

Metalloproteins are a subclass of proteins that contain one or more metal atoms and/or clusters as part of their molecular structure. Metalloproteins are found in all domains of life, including bacteria, archaea, and eukaryotes, and perform a wide range of functions. For example, hemoglobin and myoglobin transport molecular oxygen in bloodstream and muscles, respectively [1, 2]. Carbonic anhydrase, superoxide dismutase, and cytochrome c oxidase catalyze chemical reactions to support cellular metabolism and growth [3–5]. Zinc finger proteins and ceruloplasmin sense changes in the cellular environment while ferredoxins can relay electrons to various redox partners [6–8]. Several hundred different types of metalloproteins have been discovered to date, with much more remaining in obscurity.

A variety of metal atom(s) including iron, zinc, copper, and magnesium have been found within the molecular structures of metalloproteins. These and/or other metal atoms are often coordinated to specific amino acid residues within the protein, influencing conformation and function(s). The number of metal atoms within a metalloprotein can vary from one Cu atom in a plastocyanin to several thousand like Fe atoms in a ferritin [9, 10]. The local structure of metal atom(s) can also be vastly different as shown in Fig. 1. Perhaps one of the most well known is the Fe-S clusters found in the metalloproteins that play a role in electron transfer and/or redox biochemistry [11–13]. Molybdenum cofactor and copper centers are other types that differ in terms of both constituent metal atoms and structure of the cluster [9, 14].

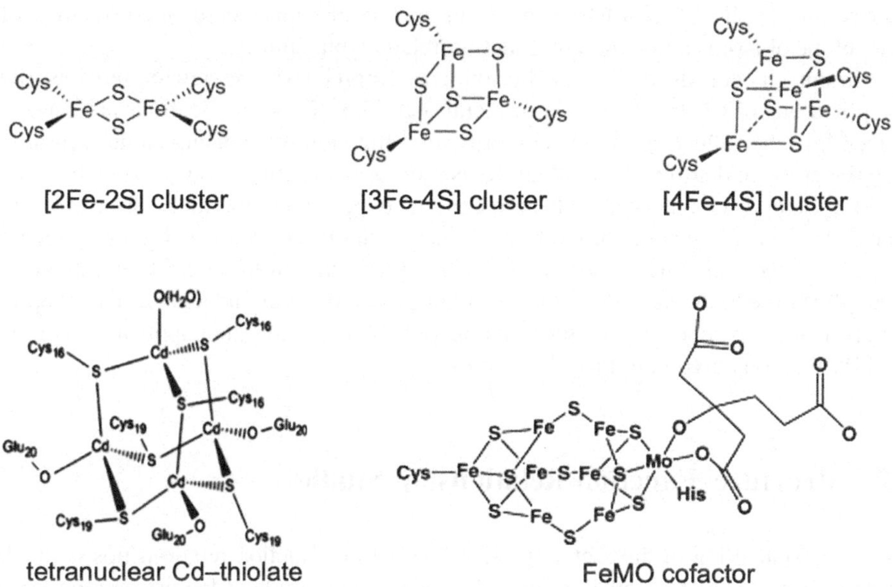

Fig. 1 Different molecular structures of metal atom(s) found in various kinds of metalloproteins

Metal atoms are incorporated into metalloproteins via a series of steps that typically take place during and/or after folding. This process involves the specific binding of metal ion(s) to amino acid residue(s) within the protein structure. The amino acids that commonly serve as ligands (binding sites) for metal ions include histidine (His), cysteine (Cys), aspartic acid (Asp), glutamic acid (Glu), and serine (Ser). The specific amino acid(s) involved in metal coordination and the arrangement of ligands around the metal ion(s) determine structure and properties of metalloproteins. Chaperones are often required as they play versatile roles ranging from metal ion acquisition and delivery to quality control of the synthesized molecules. Well-known examples are Cox17, which delivers copper ions to cytochrome c oxidase, and Hsp90 that monitors the folding and stability of client proteins, including metalloproteins [15, 16].

Much progress has been made with respect to our understanding of metal atom (s) incorporation and functional properties of metalloproteins. Cell-free protein synthesis (CFPS) methods have proven to be especially effective toward these efforts as they offer control over the composition and redox environment of the reaction buffer inside which metalloproteins are synthesized. Kuchenreuther et al., for example, elucidated the mechanism and intermediate products involved in synthesis and incorporation of Fe-Fe cluster to the hydrogenase via applying CFPS to maturation of the apo-enzyme [17, 18]. The authors also relied on the CFPS method for synthesizing and evaluating the enzymatic activity and O_2 tolerance of the metalloprotein variants in a high-throughput manner [19, 20]. Other successful applications of CFPS involving metalloproteins include incorporation of novel metal atoms and structure–function relationship studies.

In this chapter, the authors will provide a comprehensive review on the versatile applications of CFPS in the field of metalloproteins. These applications are categorized into the following three areas: structure–function relationship studies, protein engineering, and artificial metalloproteins. For each area, the emphasis is on how the CFPS methods enabled researchers to conduct experiments and make discoveries in an unprecedented manner. Literary examples from diverse fields including synthetic biology, electrochemistry, and drug discovery are chosen to demonstrate utility of the CFPS methods, as well as the resultant groundbreaking advances. The chapter ends with our perspectives on the remaining challenges and future applications of the CFPS methods involving metalloproteins.

2 Structure–Function Relationship Studies

Our understanding of the metalloproteins' structure–function relationships is much lower than the level for other types of proteins. A multitude of factors ranging from structural heterogeneity to low synthesis yields make it difficult to conduct the experimental studies on metalloproteins. The low yields, in particular, are troublesome since a highly concentrated sample is required for structural analysis including X-ray crystallography. It is difficult, however, to synthesize large amounts of metalloproteins using in vivo expression systems due to toxicity. Even when produced in large quantities inside the host cells, metalloproteins may be unstable, prone to aggregation, or difficult to purify, leading to low yields and limited availability for experimental studies [21].

CFPS methods offer the following advantages that enable one to overcome the problem of low yields and other limitations: controlled environment, isotopic labeling, and high-throughput mutagenesis. To begin with, toxicity due to a high concentration of metal ion and/or accumulation of metalloproteins is no longer an issue since it is an in vitro expression system with no intact cells. The direct access to the reaction buffer where the CFPS takes place enables precise control of the chaperones, cofactors, and reaction conditions that promote protein solubility and proper folding. CFPS methods also enable incorporation of isotopically labeled amino acids such as ^{13}C-Leu into the metalloprotein molecules. The isotopic labeling patterns can enhance the resolution and accuracy of structural data, allowing detailed analysis of metal coordination sites, ligand binding, and conformational dynamics. In addition, CFPS allows rapid, efficient production of multiple variants of metalloproteins, making it suitable for high-throughput screening assays. The assays can be used to characterize the influence of specific mutation(s) on the structure and/or function of a given metalloprotein. We will look at the examples that benefited from these advantages the CFPS methods offer in the following subsections.

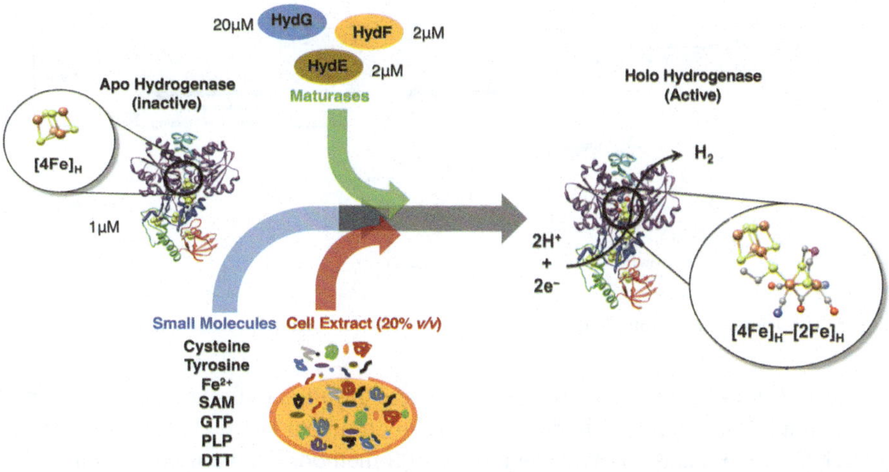

Fig. 2 An example of the Post-translational addition of metal ions during CFPS of metalloproteins

2.1 *Improved Yields*

CFPS methods can enable production of metalloproteins with high soluble titers. The traditional, in vivo expression systems often fail to achieve this due to toxicity, solubility, and/or poor cofactor assembly. These issues can be resolved by employing CFPS as it enables direct, real-time access to the environment where metalloprotein molecules are being synthesized. Toxicity, for example, is no longer an issue since the host cells are lysed before use in CFPS.

Numerous examples have been reported in the literature where CFPS allowed order(s) of magnitude improvement in soluble titers of metalloproteins. Multicopper oxidases (MCOs) are one of the examples where the soluble titer improved drastically from few tens of mg per liter to roughly 1,200 mg/L [22]. The soluble MCOs were more than 95% of the total yield and enzymatically active. Li et al. achieved this by developing an *Escherichia coli*-based cell-free (CF) transcription-translation system with the post-translational addition of copper ions (Fig. 2). The novel CF system allowed control over the buffer environment where MCOs were synthesized. Li and colleagues kept the environment dilute and filled with a high concentration of Cu^{2+}, which could not be achieved in the traditional, in vivo expression system due to the presence of cell membrane and toxicity.

Another example involves a metalloprotein with multiple, heterogeneous metal atoms, namely Cu, Zn-superoxide dismutases (hSOD1s) that consumes superoxide radicals to protect cells against oxidative stress. Ezure et al. used the CFPS method to synthesize hSOD1s and study the enzyme's molecular structure [23]. The CFPS method enabled synthesizing the enzyme molecules under the varying ratios of Cu^{2+} and Zu^{2+} over a wide range (1–400 μM). The soluble yield of enzymatically active hSOD1s varied by eightfold depending on the ratio as well as the total concentration

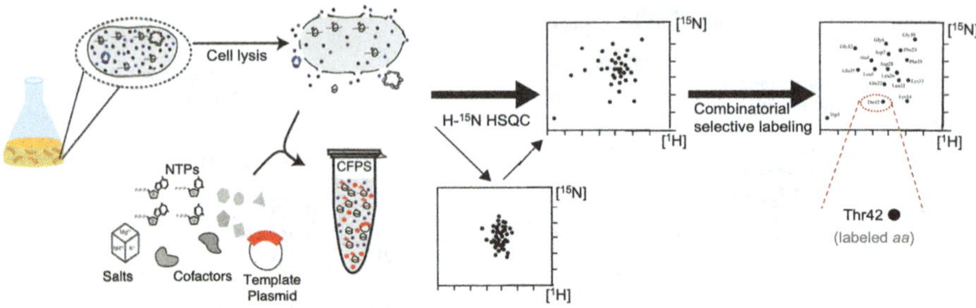

Fig. 3 Incorporation of isotopically labeled amino acids into metalloproteins via CFPS

of the two ions in the synthesis buffer. These results confirmed that an optimal concentration of metal ions in the synthesis environment is critical for the solubility and yield of the metalloproteins and that CFPS methods can be used to identify the optimal condition.

2.2 *Isotopic Labeling*

One of the difficulties associated with synthesis and characterization of metalloproteins is their susceptibility to misfolding. Previous studies suggest that as much as 80% of the total yield can be misfolded [24, 25]. To make matters worse, the proportion of the misfolded metalloprotein molecules can vary by more than two-fold even when using the same cell line and protocol. This is troublesome, especially when assessing specific activities of metalloprotein variants since one cannot determine how much variability in activity is due to changes in the proportion of misfolding versus inherent activity.

CFPS methods can help to resolve this issue by allowing one to incorporate isotopically labeled amino acids into the synthesized metalloprotein molecules (Fig. 3). The content of isotopically labeled amino acids such as ^{13}C-Leu and/or ^{15}N-Arg within the synthesized, soluble molecules can be measured as well as those in the total yield, which includes the misfolded, aggregate molecules. The labeling can thus enable one to determine specific activity of the metalloprotein variants. In fact, the author and colleagues relied on the CFPS method and isotopically labeled amino acids to characterize H_2 production activity and O_2 tolerance of a few hundred [FeFe] hydrogenase variants [20].

Another advantage the isotopic labeling by CFPS methods offers is the ability to monitor dynamics of binding at the labeled residues. Nuclear magnetic resonance (NMR) spectroscopy can be used to measure the time-course of changes in conformation of the labeled molecules, which provides insights on structural changes upon metal binding or ligand interactions. For example, Nguyen et al. incorporated ^{15}N-Ile and -Val into the specific sites of a protease and analyzed the site-specific binding of Co^{II} via NMR [26]. The results confirmed the binding of cobalt ions to the

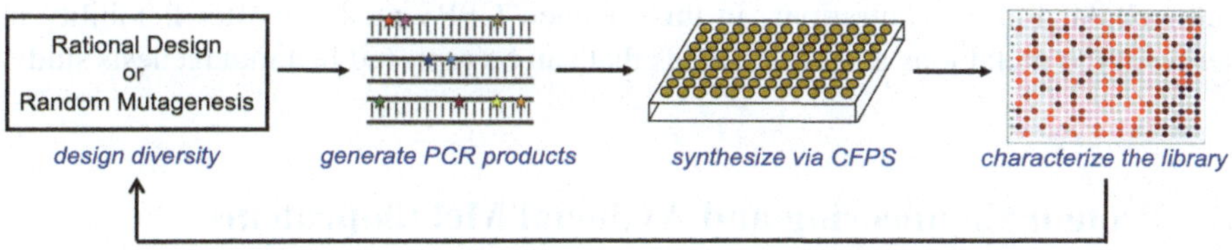

Fig. 4 High-throughput synthesis and characterization of mutagenic libraries via CFPS

designated site, as well as the resultant conformational change in the protease molecule. Bocharova and colleagues also employed the CFPS-based ^{13}C and ^{15}N labeling to study the metal-binding domain of amyloid precursor protein [27]. These examples illustrate the versatile applicability of the isotopic labeling and CFPS to studying structure–function relationships across diverse kinds of metalloproteins.

2.3 *High-Throughput Mutagenesis*

CFPS methods have proven to be efficient and effective in synthesizing a (large) library of metalloprotein variants and assessing their functional properties. Unlike in the traditional, in vivo expression systems, purification of the synthesized molecules is often unnecessary due to absence of intact membranes and relatively low concentrations of other proteins in the media. The library of variants can also be synthesized in parallel (Fig. 4) and assayed directly. This is again feasible because the reagents in the CFPS reaction mixture are dilute enough to have minimal cross-reaction with the functional assays.

The recent study conducted by the Swartz research group at Stanford University is an example where more than 10,000 variants of a metalloenzyme called [FeFe] hydrogenase were synthesized and characterized using the CFPS method [19]. By adding the random mutagenesis PCR products to the CFPS reaction mixture, the group synthesized hydrogenase variants harboring an average of 13 mutations. These variants were immediately subject to the NADPH-driven assay designed to evaluate the enzymatic activity and pH tolerance of the metalloenzyme. The results revealed an unexpected alpha helix domain that allowed up to three-fold improvement in the specific activity when replaced with a proline. Notably, the domain was a few nanometers away from the active site, which could have been easily overlooked in site-directed mutagenesis studies.

Others succeeded in analyzing the influence of metal ion type and concentration on its incorporation and ligand geometry. Nianios and colleagues studied the influence using a recombinant *Streptomyces* quercetinase as a model metalloprotein [28]. When exposed to 3 mM metal salts in the CFPS media, the quercetinase apo-enzymes were able to incorporate Ni, Co, Fe, or Mn into the active site; however, only the Ni-incorporated holo-enzymes exhibited the catalytic activity and expected ligation geometry. Such a study would have not been possible in the traditional, in vivo expression systems due to metal toxicity and/or variability in the intracellular ion concentrations. In this manner, CFPS methods offer flexibility in terms of the metal ions and amino acids that can be replaced in a mutagenesis study.

3 Protein Engineering and Artificial Metalloproteins

3.1 Protein Engineering of Metalloproteins

Protein engineering is the process of manipulating the structure and function of proteins to create the variants with enhanced or novel capabilities. It involves altering the sequence of amino acids in a protein to influence its structure and function, or combining motifs from different proteins to create entirely new ones. Protein engineering methods can be broadly divided into two categories: rational design, where modifications are based on understanding of the protein's structure and function relationship, and directed evolution, where random mutations are introduced and beneficial changes are selected and maintained. The engineered proteins have numerous potential applications in biotechnology, medicine, and materials science, ranging from enzyme catalysts in industrial processes and therapeutic proteins in medicine to biosensors in environmental monitoring.

CFPS holds tremendous potential in the field of protein engineering. CFPS allows for the direct and precise manipulation of the synthesis environment, enabling the incorporation of non-natural amino acids, isotopic labels, and/or different metal ions for the study and creation of metalloproteins. Moreover, because CFPS occurs without intact cell membranes, one can bypass the biological constraints of in vivo systems such as toxicity and the cell's own regulatory mechanisms. This allows for high-throughput production of protein variants and the synthesis of variants that might be toxic or difficult to produce in living cells. In addition, CFPS facilitates the rapid testing and optimization of engineered proteins (Fig. 4), streamlining the iterative design-build-test cycle inherent to protein engineering. The unique advantages of CFPS can accelerate discoveries and innovations in protein engineering.

The author has exploited these advantages CFPS methods offer when engineering variants of an [FeFe] hydrogenase from *C. pasteurianum* for enhanced O_2 tolerance [20, 29]. Using the CFPS method, a site-directed mutagenesis library of 278 variants having single or double amino acid replacements were synthesized in parallel and assayed. The design-build-test cycle was performed at the rate that was at least several-folds faster than the rate achievable using the traditional, in vivo system; the removal of transformation and purification steps were central to the enhanced rate. More importantly, the improved O_2 tolerance of the hydrogenase variants measured during the cycle based on the CFPS method was consistently observed using the molecules produced by the traditional, in vivo expression system.

Other researchers also employed the CFPS method for rapid engineering and discovery of the metalloprotein variants with enhanced functional properties. Another Fe-S protein, namely nitrogenase, has been successfully engineered for improved nitrogen fixation wherein the CFPS method was used to synthesize the

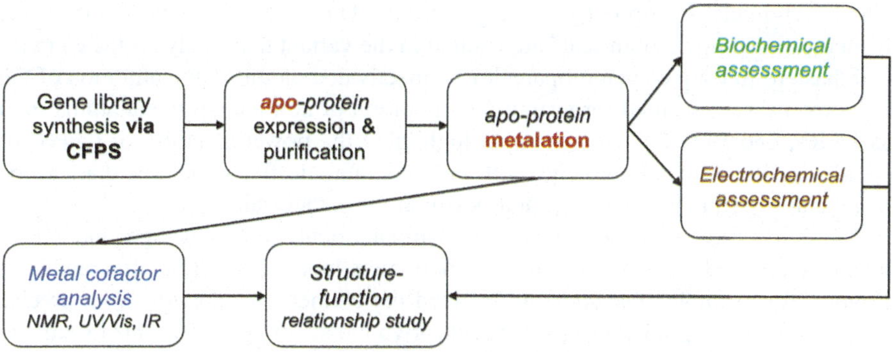

Fig. 5 Development and engineering of artificial metalloproteins via CFPS

enzyme variants [30]. The manganese peroxidase molecules with the enhanced solubility and specific activity have been discovered by applying the CFPS method as well [31]. In this manner, these and other studies underline the power and versatility of the CFPS method in driving rapid advancements in the engineering and discovery of metalloprotein variants with desirable functional attributes.

3.2 *Artificial Metalloproteins*

Artificial metalloproteins are a rapidly developing area in the field of protein engineering, in which synthetic, non-native metal centers or non-natural metal-containing prosthetic groups are introduced into proteins to confer novel functionalities. They offer exciting possibilities for the development of new catalysts, novel therapeutic agents, and advanced materials. However, producing artificial metalloproteins using conventional in vivo protein expression systems can be challenging, primarily because of the potential toxicity of some metal ions to the host cells and the difficulty of controlling precise incorporation of the metal ions.

CFPS offers the possibility to circumvent the restrictions of traditional, in vivo expression systems by providing the opportunity to directly control the synthesis environment, thus facilitating incorporation of metal ions into the protein structures. By enabling controlled co-translation of metal-containing cofactors, CFPS can promote the proper folding and function of artificial metalloproteins (Fig. 5). In addition, CFPS methods allow high-throughput screening of the synthesized protein variants, thereby accelerating the process of optimizing the structure and function of artificial metalloproteins. The potential of CFPS in this field is vast, enabling the creation of artificial metalloproteins with unprecedented properties and functionalities.

A striking example of generating artificial metalloproteins with novel functionalities using the CFPS methods can be seen in a study by Key et al. [32]. An artificial metalloprotein was synthesized by introducing Ir into the modified myoglobin, which originally incorporated a Fe-porphyrin. Directed evolution of the novel Ir-incorporated myoglobin scaffold resulted in the variant that catalyzes the enantio- and diastereoselective cyclopropanation of inactivated olefins. CF maturation of the apo-protein was essential for testing incorporation of non-natural metal atoms such as Ir, Rh, and Ru(CO). This example highlights the powerful utility of CFPS for creating artificial metalloproteins with tailored catalytic functionalities that can be beneficial in synthetic chemistry and potentially in other industries.

More recently, the combination of computational protein design and CFPS methods allowed development of de novo metalloproteins. Christoffel and colleagues, for example, designed and evolved the chimeric streptavidin with precise positioning of two gold atoms [33]. The de novo metalloenzyme was able to activate the alkyne moiety and convert into an indole carboxamide with 96% regioselectivity. The CFPS method based on the *E. coli* cell extract was utilized in the directed evolution of the computationally designed scaffold. On the other hand, Zubi et al. exploited the metal salts to control, like a switch, enzymatic activity of the de novo protein [34]. The computational method was employed to design de novo conformational switches while the CFPS method enabled rapid prototyping and high-throughput screening of the active variants. Evidently, computational protein design and CFPS methods contributed to significant advancements in the creation of de novo metalloproteins, and offered promising avenues for applications in diverse fields including bio-catalysis and sensing.

4 Challenges and Future Applications

Despite the considerable benefits offered by CFPS, there are challenges that researchers need to address when applying the methods to analyzing and/or engineering metalloproteins. The optimization of metal ion concentration in the synthesis environment remains a tricky task. While some metalloproteins require a high concentration of metal ions for their synthesis and function, too much of certain metal ions can inhibit incorporation and lower the overall yield. Another challenge lies in the formation of functional metal centers. The coordination of metal ions by the correct ligands in the protein is a complex process and varies widely among different metalloproteins. Ensuring the precise incorporation of metal atoms into the proteins and the correct formation of metal clusters is still an area where improvements can be made.

Looking ahead, the potential applications of CFPS methods for metalloprotein engineering are vast and exciting. There are still numerous metalloproteins whose structures and functions remain unexplored. With its ability to control reaction conditions and synthesize a wide range of protein variants, CFPS can greatly expedite the discovery and understanding of these proteins. In the field of synthetic biology, CFPS can be used to engineer artificial metalloproteins with desired functions, such as novel biocatalysts for sustainable chemistry, environmental remediation, or bioenergy applications.

Finally, the combination of CFPS methods with advanced analytical techniques like cryo-electron microscopy, NMR spectroscopy, and X-ray crystallography will likely revolutionize the structural and functional studies of metalloproteins. The isotopic labeling capabilities of CFPS methods, for example, will further enhance the accuracy of these techniques, allowing detailed studies of metal coordination sites, ligand binding, and conformational dynamics. As our understanding of the unique properties and diverse roles of metalloproteins in biology continues to grow, CFPS methods will remain a powerful tool at the forefront of these discoveries.

References

1. Mairbäurl H, Weber RE (2011) Oxygen transport by hemoglobin. Compr Physiol 2(2): 1463–1489
2. Wittenberg BA, Wittenberg JB (1989) Transport of oxygen in muscle. Annu Rev Physiol 51(1): 857–878
3. Lindskog S (1997) Structure and mechanism of carbonic anhydrase. Pharmacol Ther 74(1): 1–20
4. Tainer JA, Getzoff ED, Richardson JS, Richardson DC (1983) Structure and mechanism of copper, zinc superoxide dismutase. Nature 306(5940):284–287
5. Yoshikawa S, Shimada A (2015) Reaction mechanism of cytochrome c oxidase. Chem Rev 115(4):1936–1989
6. Laity JH, Lee BM, Wright PE (2001) Zinc finger proteins: new insights into structural and functional diversity. Curr Opin Struct Biol 11(1):39–46
7. Fox PL, Mukhopadhyay C, Ehrenwald E (1995) Structure, oxidant activity, and cardiovascular mechanisms of human ceruloplasmin. Life Sci 56(21):1749–1758
8. Senda M, Kishigami S, Kimura S, Fukuda M, Ishida T, Senda T (2007) Molecular mechanism of the redox-dependent interaction between NADH-dependent ferredoxin reductase and Rieske-type [2Fe-2S] ferredoxin. J Mol Biol 373(2):382–400
9. Colman PM, Freeman HC, Guss JM, Murata M, Norris VA, Ramshaw JA, Venkatappa MP (1978) X-ray crystal structure analysis of plastocyanin at 2.7 Å resolution. Nature 272(5651): 319–324
10. Crichton RR (1971) Ferritin: structure, synthesis and function. N Engl J Med 284(25): 1413–1422
11. Adman ET, Sieker LC, Jensen LH (1973) The structure of a bacterial ferredoxin. J Biol Chem 248(11):3987–3996
12. Shima S, Pilak O, Vogt S, Schick M, Stagni MS, Meyer-Klaucke W, Warkentin E, Thauer RK, Ermler U (2008) The crystal structure of [Fe]-hydrogenase reveals the geometry of the active site. Science 321(5888):572–575
13. Bilder PW, Ding H, Newcomer ME (2003) Crystal structure of the ancient, Fe– S scaffold IscA reveals a novel protein fold. Biochemistry 43(1):133–139
14. Wuebbens MM, Liu MT, Rajagopalan KV, Schindelin H (2000) Insights into molybdenum cofactor deficiency provided by the crystal structure of the molybdenum cofactor biosynthesis protein MoaC. Structure 8(7):709–718
15. Palumaa P, Kangur L, Voronova A, Sillard R (2004) Metal-binding mechanism of Cox17, a copper chaperone for cytochrome c oxidase. Biochem J 382(1):307–314
16. Young JC, Moarefi I, Hartl FU (2001) Hsp90: a specialized but essential protein-folding tool. J Cell Biol 154(2):267

17. Kuchenreuther JM, Myers WK, Stich TA, George SJ, NejatyJahromy Y, Swartz JR, Britt RD (2013) A radical intermediate in tyrosine scission to the CO and CN− ligands of FeFe hydrogenase. Science 342(6157):472–475
18. Kuchenreuther JM, Myers WK, Suess DL, Stich TA, Pelmenschikov V, Shiigi SA, Cramer SP, Swartz JR, Britt RD, George SJ (2014) The HydG enzyme generates an Fe (CO) 2 (CN) synthon in assembly of the FeFe hydrogenase H-cluster. Science 343(6169):424–427
19. Koo J, Schnabel T, Liong S, Evitt NH, Swartz JR (2017) High-throughput screening of catalytic H2 production. Angew Chem Int Ed 56(4):1012–1016
20. Koo J, Swartz JR (2018) System analysis and improved [FeFe] hydrogenase O2 tolerance suggest feasibility for photosynthetic H2 production. Metab Eng 49:21–27
21. Mounicou S, Lobinski R (2008) Challenges to metallomics and analytical chemistry solutions. Pure Appl Chem 80(12):2565–2575
22. Li J, Lawton TJ, Kostecki JS, Nisthal A, Fang J, Mayo SL, Rosenzweig AC, Jewett MC (2016) Cell-free protein synthesis enables high yielding synthesis of an active multicopper oxidase. Biotechnol J 11(2):212–218
23. Ezure T, Suzuki T, Ando E, Nishimura O, Tsunasawa S (2009) Expression of human cu, Zn-superoxide dismutase in an insect cell-free system and its structural analysis by MALDI-TOF MS. J Biotechnol 144(4):287–292
24. Boyer ME, Stapleton JA, Kuchenreuther JM, Wang CW, Swartz JR (2008) Cell-free synthesis and maturation of [FeFe] hydrogenases. Biotechnol Bioeng 99(1):59–67
25. Kuchenreuther JM, Grady-Smith CS, Bingham AS, George SJ, Cramer SP, Swartz JR (2010) High-yield expression of heterologous [FeFe] hydrogenases in Escherichia coli. PLoS One 5(11):e15491
26. Nguyen TH, Ozawa K, Stanton-Cook M, Barrow R, Huber T, Otting G (2011) Generation of Pseudo contact shifts in protein NMR spectra with a genetically encoded cobalt (II)-binding amino acid. Angew Chem Int Ed 50(3):692–694
27. Bocharova OV, Urban AS, Nadezhdin KD, Bocharov EV, Arseniev AS (2013) Bacterial and cell-free production of APP671-726 containing amyloid precursor protein transmembrane and metal-binding domains. Biochemistry 78:1263–1271
28. Nianios D, Thierbach S, Steimer L, Lulchev P, Klostermeier D, Fetzner S (2015) Nickel quercetinase, a "promiscuous" metalloenzyme: metal incorporation and metal ligand substitution studies. BMC Biochem 16:1–11
29. Koo J, Shiigi S, Rohovie M, Mehta K, Swartz JR (2016) Characterization of [FeFe] hydrogenase O2 sensitivity using a new, physiological approach. J Biol Chem 291(41):21563–21570
30. Payá-Tormo L, Coroian D, Martín-Muñoz S, Badalyan A, Green RT, Veldhuizen M, Jiang X, López-Torrejón G, Balk J, Seefeldt LC, Burén S (2022) A colorimetric method to measure in vitro nitrogenase functionality for engineering nitrogen fixation. Sci Rep 12(1):10367
31. Ninomiya R, Zhu B, Kojima T, Iwasaki Y, Nakano H (2014) Role of disulfide bond isomerase DsbC, calcium ions, and hemin in cell-free protein synthesis of active manganese peroxidase isolated from Phanerochaete chrysosporium. J Biosci Bioeng 117(5):652–657
32. Key HM, Dydio P, Clark DS, Hartwig JF (2016) Abiological catalysis by artificial haem proteins containing noble metals in place of iron. Nature 534(7608):534–537
33. Christoffel F, Igareta NV, Pellizzoni MM, Tiessler-Sala L, Lozhkin B, Spiess DC, Lledos A, Marechal JD, Peterson RL, Ward TR (2021) Design and evolution of chimeric streptavidin for protein-enabled dual gold catalysis. Nature Catalysis 4(8):643–653
34. Zubi YS, Seki K, Li Y, Hunt AC, Liu B, Roux B, Jewett MC, Lewis JC (2022) Metal-responsive regulation of enzyme catalysis using genetically encoded chemical switches. Nat Commun 13(1):1864

Adv Biochem Eng Biotechnol (2023) 185: 59–90
https://doi.org/10.1007/10_2023_227

Published online: 13 June 2023

Cell-Free Display Techniques for Protein Evolution

Jiaojiao Li, Youhui Yang, Jinjin Li, Peixian Li, and Hao Qi

Contents

Abstract Cell-free protein synthesis (CFPS) with flexibility and controllability can provide a powerful platform for high-throughput screening of biomolecules, especially in the evolution of peptides or proteins. In this chapter, the emerging strategies for enhancing the protein expression level using different source strains, energy systems, and template designs in constructing CFPS systems are summarized and discussed in detail. In addition, we provide an overview of the ribosome display, mRNA display, cDNA display, and CIS display in vitro display technologies, which

J. Li, Y. Yang, J. Li, and P. Li
School of Chemical Engineering and Technology, Tianjin University, Tianjin, China

H. Qi (✉)
School of Chemical Engineering and Technology, Tianjin University, Tianjin, China

Frontiers Science Center for Synthetic Biology (Ministry of Education), Tianjin University, Tianjin, China
e-mail: haoq@tju.edu.cn

can couple genotype and phenotype by forming fusion complexes. Moreover, we point out the trend that improving the protein yields of CFPS itself can offer more favorable conditions for maintaining library diversity and display efficiency. It is hoped that the novel CFPS system can accelerate the development of protein evolution in biotechnological and medical applications.

Graphical Abstract

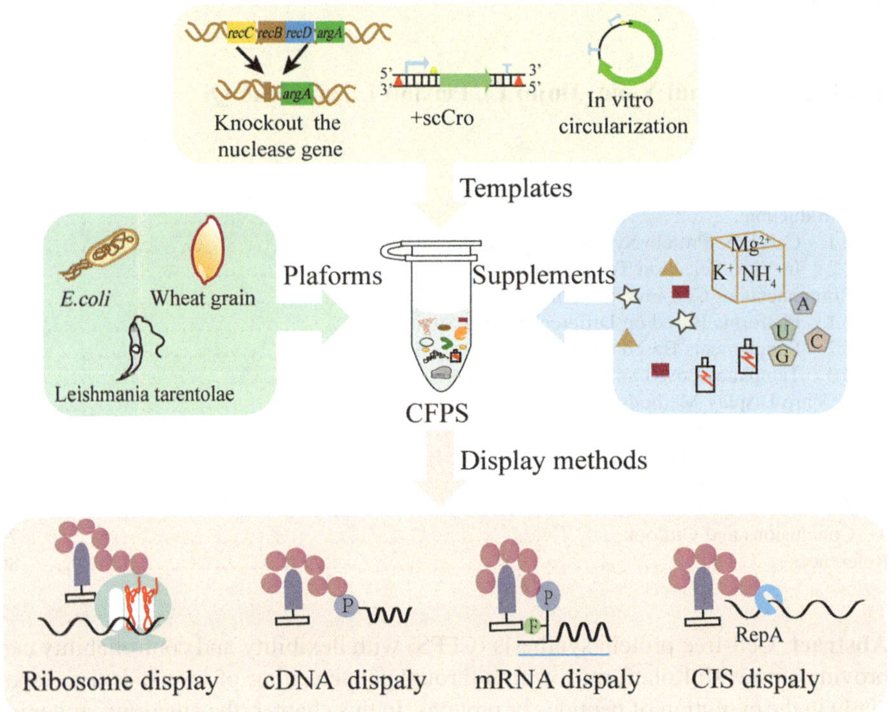

Keywords cDNA display, Cell-free system, CIS display, mRNA display, Protein evolution, Ribosome display

1 Introduction

1.1 Cell-Free Protein Synthesis Systems

Cell-free synthetic biology expanded to various biological fields, enabling the powerful engineering of biological functions. Cell-free protein synthesis (CFPS)

Table 1 A comparison of cell-free and in vivo protein synthesis systems

Features	In vitro cell-free system	In vivo cell-based system
Type of DNA template	PCR products or plasmids	Plasmids
Design-build-test cycles	1–2 days	1–2 weeks
Cost	Relatively low	Relatively high
Post-transcriptional modifications	Hard	Easy
Genetic manipulation	Easy	Hard
High-throughput biochemical production	Easy	Hard
Synthesis of membrane proteins and toxic proteins	Easy	Hard
Gene circuit testing	Simple	Complex
Incorporation of unnatural amino acids	Easy	Hard

systems provide an open platform that can be directly manipulated without the bottlenecks of membrane-bound barriers in living cells. It has emerged as a key tool for the vitro synthesis of challenging proteins, including antibodies [1, 2], membrane proteins [3–5], and large proteins [6, 7]. At present, cell-free systems have been classified into two major types. The first, extract-based system is composed of a cell lysate, supplemented with the essential transcription and translation elements, cofactors and substrates, energy and salts, as well as the translation template encoding the protein of interest [8]. The second, so-called protein synthesis using recombinant elements (PURE) system, relies on protein synthesis using a toolbox of recombinant factors purified via histidine (His)-tags [9, 10]. Considering the labor-intensity aspect of this production process, new emerging systems are seeking a simpler, more flexible, and cost-effective ways of protein synthesis, such as the "ePURE" [11], "TraMOS PURE" [12], "PURE 3.0" [13], and "OnePot PURE" system [14]. Although this PURE system has been continuously optimized, there are still some limitations in building a fully-characterized PURE system that can be quickly broadened to various fields. Considering the high cost and the tedious preparation in PURE system, a system based on a cell-free extract can be more widely and feasibly used in laboratory research and industrial production. Moreover, the CFPS platform with its open nature can offer more flexibility and controllability than living cells. Here, we present an overview and comparison of the features related to the processes of protein preparation, manipulation, and production between the in vitro and in vivo systems (Table 1).

1.2 In Vitro Selection Technologies

Directed evolution of biomolecules has been applied in the selection of targeting peptides and proteins with desired properties. During the past two decades, phage display [15, 16] and yeast-surface display [17] have been widely used for the selection of functional enzymes and high-affinity antibodies from random libraries. However, these methods have some inherent drawbacks, such as low transform efficiency, small library size (range of 10^9 to 10^{10}), and complex cloning operations in a cell-based system. To overcome the limitations associated with cell-based display systems, several in vitro display techniques have been developed for the high-throughput evolution and screening of biomolecules based on cell-free protein synthesis systems [18]. Considering the powerful advantages of the in vitro approach discussed above, these selection methods, including ribosome display, mRNA display, cDNA display, CIS display, and other display systems took advantage of the ability to couple genotypes with phenotypes by linking the mRNA to its encoded peptide during the translation process. Although each of the different display techniques mentioned above has its specific characteristics, they all share two advantages: the controllability of designing a large, highly diverse library, and the simplicity of evolving protein molecules through the iterative screening of random libraries. In fact, in vitro display platforms have been successfully applied in a wide variety of fields, such as the discovery of antibody drugs [19], selection of small molecules [20], and the directed evolution of enzymes [21].

This chapter presents a comprehensive understanding of high-throughput screening for protein evolution by focusing on important factors in the CFPS system with an enhanced scope of variation for in vitro display systems. Firstly, three pivotal factors in the construction of the CFPS system are reviewed, including different source strains, energy systems, and template designs. Additionally, we outline improved variation for the in vitro selection of functional peptides, mainly focusing on ribosome display, mRNA display, cDNA display, and CIS display. Finally, we summarize the advantages and disadvantages of four different display systems as well as their further improvement and novel applications in protein evolution.

2 Components of Cell-Free Protein Synthesis System

To achieve the ultimate goal of low cost and high efficiency, the construction process of the CFPS platform for recombinant protein production has been gradually improved. In the following chapter, details of the main factors crucial for the construction process will be discussed, including the platforms based on different strains, optimization of the reaction components, as well as different template designs.

2.1 *Platforms Based on Different Source Strains*

CFPS technology can harness all cellular resources to synthesize a single protein. In the last two decades, various platforms derived from different strain extracts have emerged for specific target proteins [22]. In general, the sources for the construction

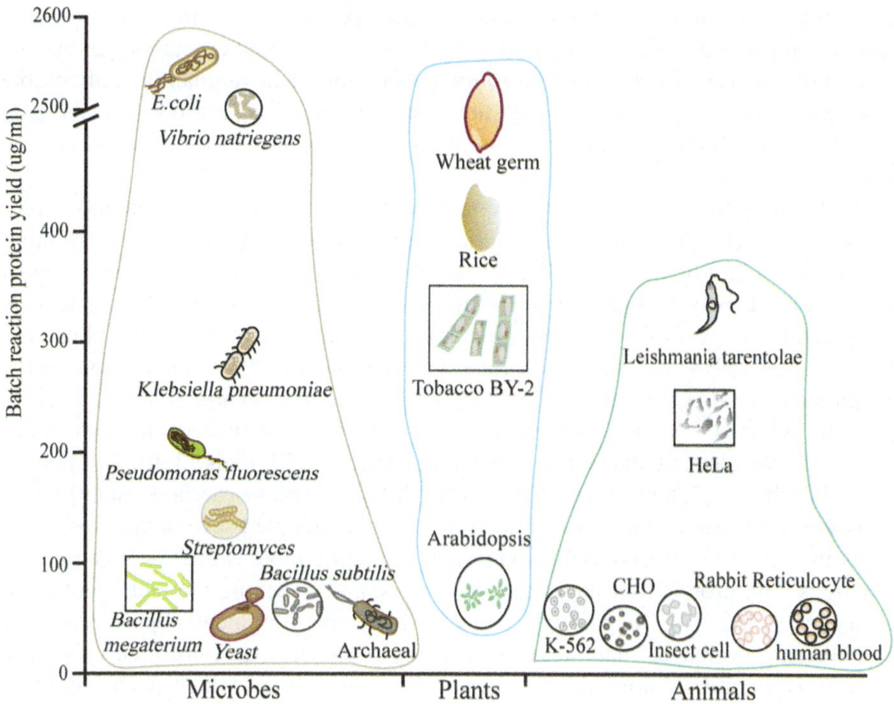

Fig. 1 Emerging platforms based on different source organisms for the constructed CFPS systems. The presented platforms based on microbes, plant- or animal cells show different levels of in vitro green fluorescent protein (GFP) expression

of CFPS platforms are divided into microbes, plants, and animals according (Fig. 1). Commonly used microorganisms mainly include *Escherichia coli*, *Streptomyces* sp., *Vibrio natriegens*, archaea, and yeasts. The other two main host source types are mainly plant and animal cells.

As a workhorse model organism, *E. coli* was the first widely used source of the bacterial S30 extract for CFPS system construction in synthetic biology applications [23]. Both B- and K-type strains of *E. coli* were investigated to develop a generalizable method of preparing crude extracts through sonication with similar productivity [24]. Crude extracts of *E. coli* were widely adopted due to a number of advantages. First, *E. coli* allows a low-cost and fast extract preparation process, which helped the emergence of economical strategies. For example, a new ATP regeneration system based on maltose can synthesize up to 2.3 g/L of a reporter protein in batch-mode reactions [25]. The addition of lactose led to a 188% increase in protein yield when maltodextrin was replaced by PEP as an energy source, achieving ultralow cost [26]. Second, it is scalable and flexible to adopt the different reaction formats for high protein yields. This open cell-free system can be extended to 100 L volumes, yielding 0.7 g/L granulocyte-macrophage colony-stimulating factor (rhGM-CSF) in standard bioreactors [27]. More importantly, it enables the industrial production of disulfide-bonded biotherapeutic proteins in vitro. Nevertheless, *E. coli* S30 extracts are commonly used with T7 promoter and exogenous T7 RNA polymerase, which is limited to the production of large quantities of soluble proteins at low temperatures. In addition, many proteins require post-translational modification, which cannot be successfully carried out using the reported cell-free systems based on *E. coli.*

To broaden the spectrum of source strains for CFPS, the preparation and exploration of extracts based on novel bacteria might offer alternatives for circumventing the disadvantages of the *E. coli* S30 extract system. The S30 extract of *Pseudomonas fluorescens* displayed high efficiency of His6-tagged GFP synthesis over a wide temperature range (4–37°C) [28]. Importantly, characterization of the ribosome-binding sites (RBS) has been combined with the batch CFPS system based on the *P. fluorescens* ATCC12633 strain [29]. A *Klebsiella pneumoniae*-derived CFPS system with a deletion of the *wzy* gene was optimized to synthesize 253 ± 15.79 μg/mL of sfGFP and expand the current CFPS toolkit [30]. To improve the solubility of proteins expressed from high GC genes, researchers established a *Streptomyces lividans*-based cell-free system, which adopted several effective measures to increase synthesis yields [31], such as adding more extract, supplementing reactions with protein translation-related factors, and screening more efficient regulatory parts [32–34]. In particular, this system can offer new opportunities for synthesizing valuable natural products such as the nonribosomal peptides, tambromycin and valinomycin. Lysates of *Bacillus subtilis*, which is an important soil organism, can significantly increase the yield of GFPmut3b to 0.8 μM in robust cell-free reactions over several hours [35–37]. Another unique system derived from archaea was published which synthesizes the thermostable proteins active at high temperatures, including *Thermus thermophilus*, *Thermococcus kodakaraensis*, and *Sulfolobus solfataricus* [38–40]. *Vibrio natriegens*, a marine bacterium with a doubling time of 9.8 min 60% more ribosomes/cell than *E. coli*, is an exciting host for establishing a new cell-free system[41, 42]. An emerging simplified preparation protocol based on *V. natriegens* crude extract was reported to increase the titer of sfGFP to 1.6 g/L [43]. The described *V. natriegens* CFPS system is productive and simple to use as a platform to broaden new areas in synthetic biology [44].

Budding yeast (*Saccharomyces cerevisiae*), another popular source of components for cell-free systems, can produce complex modified proteins folded correctly [45]. Cell extracts based on *S. cerevisiae* offer large-scale preparation at low cost and show high stability during long-term storage. However, this yeast cell-free system also suffered from low translation efficiency (in comparison with *E. coli*). To circumvent this drawback, several alternative strategies emerged and were optimized for increasing the protein synthesis rate, including the optimization of extract preparation and reaction conditions [46], substrate replenishment and by-product removal [47], the disruption of ribosome-inactivating factors [48], and template engineering [49]. Moreover, considering the requirement for the expensive energy carriers PEP and phosphocreatine, research has gradually focused on the development of a novel energy regeneration system, which is expected to develop into a platform for industrial production strains in the future. The demonstration of using glucose and phosphate as the substrates to regenerate ATP for the synthesis of active luciferase confirmed the feasibility of alternative approaches, despite the fact of low protein yields [50]. Recently, further optimization of the combination of the in vivo and in vitro genetic metabolic engineering resulted in an increase of 2,3-butanediol titers and volumetric productivities, expanding its application potential to cell-free prototyping and biomanufacturing [51].

At present, a variety of CFPS based on plants cell extracts are also available. The wheat germ extract (WGE), one of the most commonly used among plants cell-free protein expression systems, can be customized for applications in various research areas including the soluble expression of high-quality proteins, the characterization of specific monoclonal antibodies, and the exploration of structural proteomics [52–54]. Since the dormant embryo contains rich protein factors and ribosomes, the wheat germ system is favorable for long biochemical studies and small-scale protein expression. The novel protocols described here are emerging to optimize the yields and activity of target antibodies and malaria vaccine candidates, currently in particular for high-throughput immuno-screening approaches [55, 56]. Nevertheless, there are still some functions in the system that need to be improved for expanding its commercial applications, such as long preparation time, as well as no ability to introduce protein glycosylation. Compared with the WGE, the demand for eukaryotic systems with high production efficiency and rapid preparation is imperative for large-scale applications. The highly-productive extract derived from tobacco Bright Yellow 2 (BY-2) heterotrophic cell culture plastids significantly shortened the preparation time to 4–5 h [57, 58]. The reported rice callus extract (RCE) system based on tissue culture was optimized to not only produce sub-g/L titers of sfGFP and mCherry, but also synthesize the large membrane protein hERG [59]. In addition, a cell-free extract was also obtained from Arabidopsis and served as a favorable platform for elucidating the relevant genetic mechanisms, especially in the degradation and stability of mRNAs [60], and the measurement of DNA base damage [61].

Given the nature of eukaryotic proteins, a CFPS system based on animal cell extract is beneficial for their production, folding, and modification. The rabbit reticulocyte lysate (RRL) system has been successfully demonstrated to express Hepatitis B virus capsid protein and assemble it into capsids under native conditions [62]. To break the restriction of low protein synthesis rates, the recently emerged hybrid RRL system based on energetic metabolism was combined with other mammalian ribosome or viral enhancers of translation to sustain long-lasting translation for increasing yields [63, 64]. Currently, cell-free systems based on cultured Chinese hamster ovary (CHO) cells are mainly used for the production of active target proteins, such as glycosylated and integral membrane proteins [65–67]. However, its characteristics are similar to the CFPS system based on HeLa cell [68], K562 cell [69], and human blood [70], which feature authentic post-translational modifications and natural codon usage. Various strategies have been focused on the improvement of translation initiation, such as the addition of translation factors [71] and designed templates containing IRES sequences [72]. The CHO system and human cell lines are promising platforms for industrial protein production, and more importantly, commercial kits have been developed and made available for research.

The use of eukaryotic cell-free systems to synthesize proteins with post-translational modifications is becoming increasingly prominent. An insect-based system platform has been constructed using *Spodoptera frugiperda* (*Sf* 21) cells, which contain endogenous microsomes that promote correct protein folding and covalent modification. Due to the properties of the insect system, the developed preparation methods can be used to efficiently produce antibodies [73], ubiquitin-conjugated proteins [74], and AB5 toxins [66]. What's more, it is also a powerful platform for studying the structure and interaction of mutated proteins, which can be combined with the insertion of non-canonical amino acids (ncAAs) using bio-orthogonal systems [75–77]. Although the system may produce a limited yield of target protein due to the generation of by-products, the yield is fully able to meet the requirements for the structural characterization of proteins with post-translational modifications. It offers a great advantage in expressing heterologous proteins, especially in the study of the mechanisms of action of novel toxins. Recently, the *Leishmania tarentolae* cell-free protein expression (LTE) system was adopted for rapid screening and evaluation of interactions between proteins [78–80]. One of the advantages of this LTE system is the ability to combine the transcription–translation system and engineered protein templates in multi-well plate format [81]. A variety of cheap and user-friendly platforms have sprung up based on the needs of different fields, including the construction of cost-effective protein arrays [82].

2.2 Supplements Based on Different Energy Sources

Cell extract, supplements, and templates are the three basic components of CFPS which work cooperatively to sustain the whole process of metabolism, transcription, and translation. It is important to note that the physicochemical interactions of supplements play an unpredictable role in the complex system, despite most literature focusing on the advanced application of various platforms. The different supplements added to CFPS reactions evolved into major prominent systems based on their secondary energy sources (Fig. 2). The original creatine phosphate

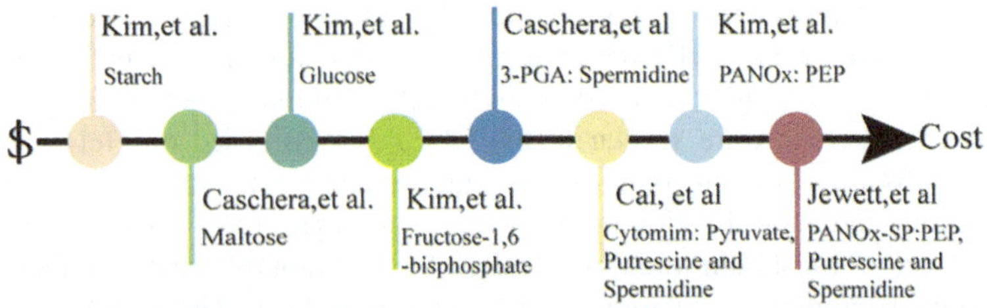

Fig. 2 The different supplements added to CFPS reactions evolved into major prominent systems based on secondary energy sources

and creatine kinase system could be used as an energy regeneration system for chloramphenicol acetyltransferase (CAT) protein synthesis [83]. In 2001, the Swartz group made the initial research to prolong the CFPS reaction by proposing the "PANOx" system, which contains phosphoenolpyruvate (PEP), amino acids, nicotinamide dinucleotide (NAD), and oxalic acid [84]. Lately, the improved PANOx-SP system with the addition of spermidine and putrescine was developed [85]. Variations of the slightly modified PANOx-SP systems have emerged due to the cost of reducing reagent. In addition, an alternative based on the oxidative phosphorylation for the supply of energy was developed in the Cytomim system, which mimicked the cytoplasmic environment and removed the expensive PEP and exogenous enzymes. These cell-free systems use reoptimized protocols for simplified operation and reduced cost in the laboratory, despite the requirement for a glucose-rich medium [24, 86, 87]. In addition, the 3-phosphoglycerate (3-PGA) system combined with Gam protein in CFPS was reported by the Chatterjee lab, which displays more stable protein synthesis ability than PEP [88]. It therefore stands to reason to increase protein yields by constructing new metabolic pathways using alternative energy sources, such as fructose-1,6-bisphosphate (FBP), glucose, maltose, and starch [25, 89, 90]. Besides the energy source, it is essential to select the best candidates for the other key components of the CFPS reaction. Various tests with different cofactors [91], significant cations (ionic compounds) [92], and crowding agents [93] were further conducted to increase the expression efficiency. Overall, manipulating the choice of supplements in CFPS offers more flexibility for different research purposes because of the types, cost, efficiency, and yield of target proteins. There is still much room to improve and optimize the suitable protocols for a particular protein. These emerging energy alternative systems revealed the future trend of low cost and high yields to promote the development of protein evolution in high-throughput way.

2.3 Templates Based on Different Designs

In CFPS, the different form of the target protein template also plays an important role in improving the protein yield. The added template was generally based on the plasmids, linear expression templates (LETs), and mRNA. All of the approaches described above still failed to generate protein yields from LETs comparable to those from plasmids, due to degradation by native nucleases in the cell extract. Although plasmids were widely used in CFPS, their construction takes a few days and is a laborious process, which consists of synthesis, cloning, and isolation, while the preparation of the LETs via PCR takes only a few hours. To some extent, the directly utilized LETs can expedite the research and screening of protein libraries in a high-throughput manner. Therefore, current methods of the genome engineering, the addition of nuclease inhibitors and the design of LETs itself are summarized for improving protein expression in protein evolution.

To extend the half-life of the LETs in CFPS, the most beneficial measure is to attempt to remove the effects of nuclease activity (Fig. 3a). Researchers modified the *E. coli* genome and added a streptavidin-binding peptide (SBP) tag at the 3′ terminus of the target gene, including the A19Δ*recD*Δ*endA* mutant [94] and RecD-SBP strain [95], enabling the removal of the exonuclease activity. Especially, the recently reported Δ*recBCD* extracts of *E. coli* BL21 supplemented with specifically optimized buffers enable productivities close to plasmid-driven protein expression [96].

Given the complexity of genomic alteration, multiple advances to inhibit the native exonucleases are emerging as alternatives for improving the yields of LETs. Three strategies based on nuclease inhibitors have been reported for protecting the LETs from degradation (Fig. 3b). The initially supplemented protein, GamS, an inhibitor of RecBCD, stabilizes LETs in *E. coli* extracts and produces a twofold amount of protein compared to the extracts without GamS [88]. Another promising method is the utilization of the six repeated crossover hotspot instigator DNA (Chi6 DNA) in CFPS for enhancing the stability of LETs [97]. Previous studies have reported that Chi DNA, the short DNA motif (5′-GCTGGTGG-3′), was stalled by the RecBCD [98]. Therefore, more RecBCD proteins have been recognized by the Chi DNA, and more LETs are used for increasing the yield, exhibiting great performance from no detectable level to near 23% of the yield obtained with the plasmid template. In addition, selecting the most potent small organic molecule to reduce the AddAB and RecBCD helicase-nucleases activity, such as the CID 697851 and CID 1517823, is also an alternative method for achieving linear DNA protection [99]. In general, the main strategy summarized above is to significantly improve protein expression levels by adding beneficial components or removing harmful components in the reaction environment of linearized templates.

To promote the development of the LETs-based CFPS system, various studies not only concentrated on the choice of the selected nuclease inhibitors, but also proposed the potential to enhance their stability by redesigning the template structure (Fig. 3c). At present, researchers mainly focus on improving their stability, which can be summarized into the following three strategies. The first strategy is to incorporate sequences of DNA-binding proteins at both termini of the template, which prevents the nuclease from degrading the template, such as the Ku [100] and scCro proteins [101], resulting in increased protein production to varying degrease. In the second method, LETs with particular chemical modifications might reduce the susceptibility to degradation using commercially available enzyme and PCR primers. One coincidental pattern of methylation by dam methyltransferase in LETs can mimic the genomic DNA and improve the expression levels by approximately 30% compared to templates without the methylation [101]. Alternative LETs with different amounts of phosphorothioates (PT) demonstrated that two PTs afforded a slight increase in yields [102]. Nonetheless, the limitation of methylation efficiency and the slight advantage of PTs can stimulate us to explore the related properties of other functional groups in improving the yield of linearized templates through chemical modification. Another alternative approach is to increase the fragment length or cyclize the template [103, 104]. This strategy increases the complexity of the secondary structure of the mRNA, which extends its half-life to some extent.

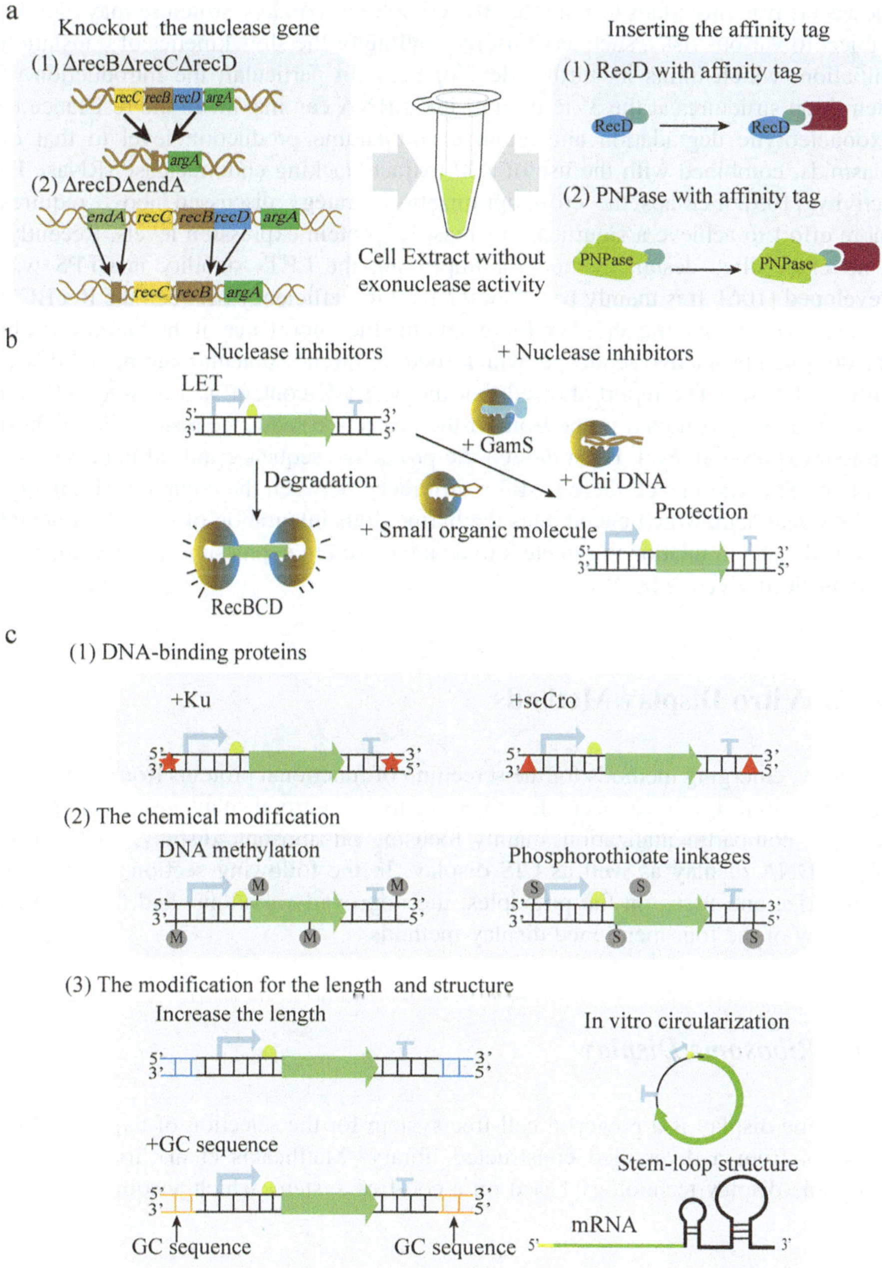

Fig. 3 Different strategies for improving protein expression from LETs. (**a**) Engineering the source organism's genome. The organisms used to generate the cell extracts can be genetically modified to remove nucleases by the deletion of nuclease genes or the genomic fusion with SBP tags for subsequent removal. (**b**) The addition of nuclease inhibitors. Three strategies are used for inhibiting the activity of native exonucleases, including the GamS protein, the Chi6 DNA as well as small

However, it is important to note that the complex secondary structure may also be subject to various risks, such as ribosome stalling or the slow kinetics of translation initiation, which limits its further development. In particular, the introduction of stem-loop structures at the 3′-termini of the mRNA can maximize the resistance to exonucleolytic degradation and attain a comparable production level to that of plasmids, combined with the use of S30 extracts lacking endonuclease (RNase E) activity [105]. Perhaps the cross-optimization strategy discussed above requires more effort to achieve a significant increase of protein expression levels. Recently, a novel in silico design method for improving the LETs stability in CFPS was developed [106]. It is mainly based on the docking efficiency between the RecBCD DNase complex and the template for evaluating the importance of the GC content in the designed protective sequence, which focused on the content, pattern, and distribution of bases. The report showed that the best GC content was between 60 and 65%, and GC distributed in the front of the protective sequence resulted in the best protein expression level. Even though the protective sequence indeed improves the protein expression level, there is still a difference between the computer simulation and the real data, which encourages the harmonious integration of the experimental protocols and simulation parameters to achieve protein expression levels compared to plasmid-derived templates.

3 In Vitro Display Methods

Recently, emerging methods for the screening of functional proteins from very large sequence libraries have been described using in vitro display systems without physical compartmentalization, mainly focusing on ribosome display, mRNA display, cDNA display as well as CIS display. In the following section, we mainly summarize and focus on the principles, the improved variations, and the research tendency of the four mentioned display methods.

3.1 Ribosome Display

Ribosome display is a powerful cell-free system for the selection of target-binding peptides from a diversified constructed library. Mattheakis et al. first proposed ribosome display technology based on a cell-free system, which accomplished the

---

Fig. 3 (continued) potential organic molecules. (**c**) The variation of LET sequence and structure. Several methods are summarized for enhancing the stability of LETs, including the incorporation of sequences for DNA-binding proteins, chemical modification, length extension, or the introduction of stem-loop structures

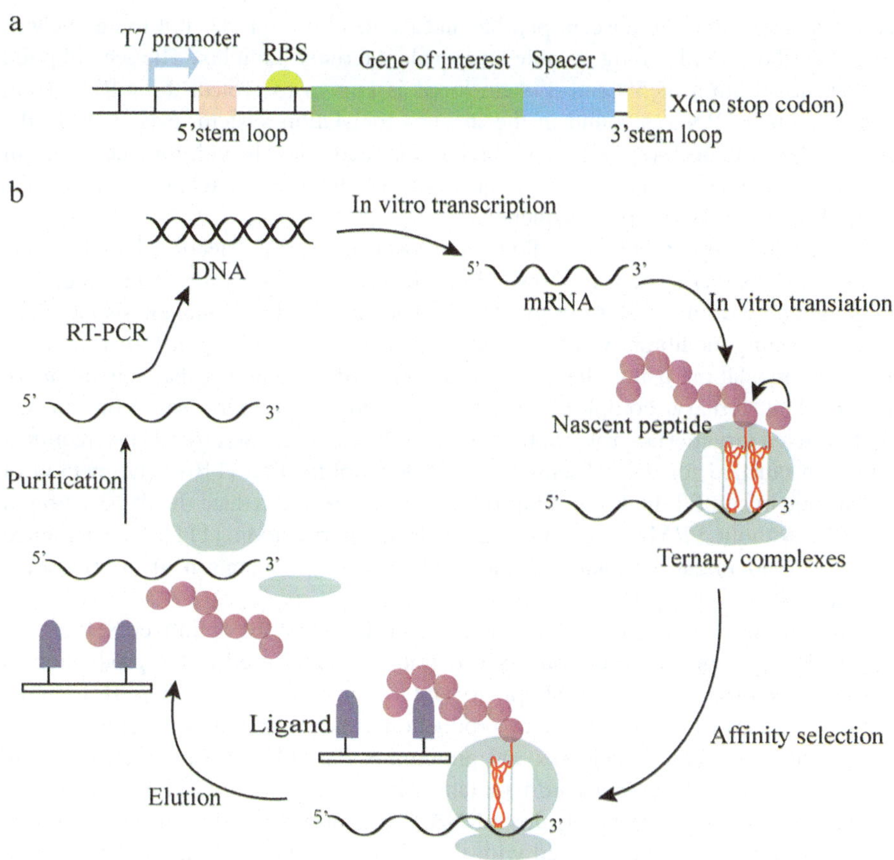

Fig. 4 Schematic representation of a ribosome display selection round. (**a**) The target gene cassette is used for ribosome display. The constructed DNA sequence consists of a T7 promoter, a 5′-stem loop, a ribosome-binding site (RBS), the sequence of interest gene, a spacer, and a 3′-stem loop. (**b**) Ribosome display cycle. This large library with the elimination of stop codon is constructed by the polymerase chain reaction (PCR). And the mRNA was transcribed and attached to the ribosome. Subsequently, the ternary protein–ribosome–mRNA (PRM) complexes were formed for affinity selection on the immobilized target. A reverse transcription reaction followed by PCR for analysis of the selected sequences

physical link between individual nascent proteins (phenotypes) to their corresponding mRNA (genotypes) [107]. Its core is characterized by the formation of stable protein–ribosome–mRNA (PRM) complexes, which can be maintained at low temperatures and elevated magnesium concentrations for up to several days [108]. As shown in (Fig. 4a), the target gene cassette mainly consists of six functional regions. More precisely, the constructed DNA sequence consists of a T7 promoter, a 5′-stem loop, a ribosome-binding site (RBS), the coding sequence of the peptide of interest, a spacer, and a 3′-stem loop. There are three key points to be noted when designing a sequence. Firstly, the elimination of the stop codons on the mRNA sequence avoids the binding and catalytic action of the release factors (RFs) and guarantees that the nascent peptide and its encoding mRNA fail to be released from the ribosome, forming a stable noncovalent ternary complex. The second point is that the addition of 5′- and 3′-stem loops is known to protect the mRNA from degradation by RNases found in the in vitro translation system, especially in the *E. coli* ribosome display system with high RNase activities. In addition, the choice of the spacer sequence and its length can influence display efficiency, including the gene III and the SecM arrest sequence.

A detailed step-by-step workflow of ribosome display selection based on the *E. coli* S30 extract system is shown (Fig. 4b) [109]. As has often been noted, the large diversity of libraries is designed by hotspot and random mutagenesis methods. DNA encoding the library is transcribed in vitro, and the resulting mRNA is used for in vitro translation using the cell-free system. Subsequently, the generation of noncovalent ternary complexes (PRM) is used for affinity selection on the immobilized target. The prokaryotic mRNA in PRM is released by EDTA treatment and then purified for the subsequent RT-PCR amplification [110]. The method of eukaryotic in situ RT-PCR can simplify the procedure and reduce the degradation of mRNA, without PRM dissociation and mRNA purification [111]. The enriched pools of binders can be analyzed and used for the next iteration of the ribosome display.

The current ribosome display methodology is based on the optimized S30 extract and PURE system. The traditional S30 extract is widely used in the production of peptides and proteins due to its simplicity and low cost of its preparation. However, a major drawback of this system is the production of target proteins of various lengths due to the presence of nucleases and proteinase. Partially RNase-deficient *E. coli* strains, such as A19Δ*recD*Δ*endA* mutant, can be used to solve the issue of low display efficiency by reducing the mRNA degradation. In addition, a eukaryotic ribosome display system coupled with the rabbit reticulocyte lysate (RRL) system was used to generate an antibody-ribosome-mRNA (ARM) complex, which could be enriched by a factor of 10^4-fold in a single display cycle [112]. Another alternative PURE system without the nuclease activity demonstrated higher cDNA yields and integrity over the S30 extract in the recovery of the selection process [113]. The favorable selection conditions for this system were attributed to the reduced abortive energy consumption and the improved mRNA stability and quality of the displayed proteins. PURE ribosome display with high controllability has been applied in the selection of monoclonal antibodies from large libraries [114, 115], as well as the epitope mapping and affinity maturation of antibodies [116].

Considerable efforts were made to optimize the key parameters of ribosome display, such as the extended incubation time for the selection of stronger binders, reaching pM affinity [117], as well as the suitable DTT concentration for the integrity of the functional targeted proteins [118]. Since ribosome display overcomes the issues of cytotoxicity and the bottleneck of low transformation efficiency in cell-based systems, it provides a powerful platform for the application of the selected functional proteins such as the high-affinity ligand-binding peptides, antibodies, and other binding proteins [119]. For example, the selected EpCAM-specific DARPin binder integrated with ribosome display can serve as a diagnostic reagent with high specificity ($K_d = 68$ pM) [120]. Another reported selected scFv with $3.74 \times 10^8\ M^{-1}$ affinity may provide a candidate reagent for the diagnosis of tuberculosis [110]. It is worth mentioning that a novel strategy was proposed for ribosome synthesis and evolution (RISE), expanding the application scope for exploring the fundamental biochemistry of ribosomes [121].

3.2 *mRNA Display*

The first description of the mRNA-display system was reported by Roberts et al. in 1997 [122]. The mRNA-display technology has been developed into a general and powerful platform for the selection of peptides with desired characteristics. In this approach, a single selection cycle mainly covers the steps shown in Fig. 5 [123]. The purified mRNA template is obtained by in vitro transcription and purification after DNase degradation, reducing the background of undesired proteins. The resulting mRNA is then covalently conjugated with a synthetic oligonucleotide containing a 3′ puromycin moiety attached by various conjugation methods, such as psoralen-mediated UV crosslinking [124] or enzymatic ligation [125]. The formation of mRNA-protein fusions can be efficiently accomplished in the RRL, or via the low activity of the nuclease excision system when the ribosome is stalled at the junction

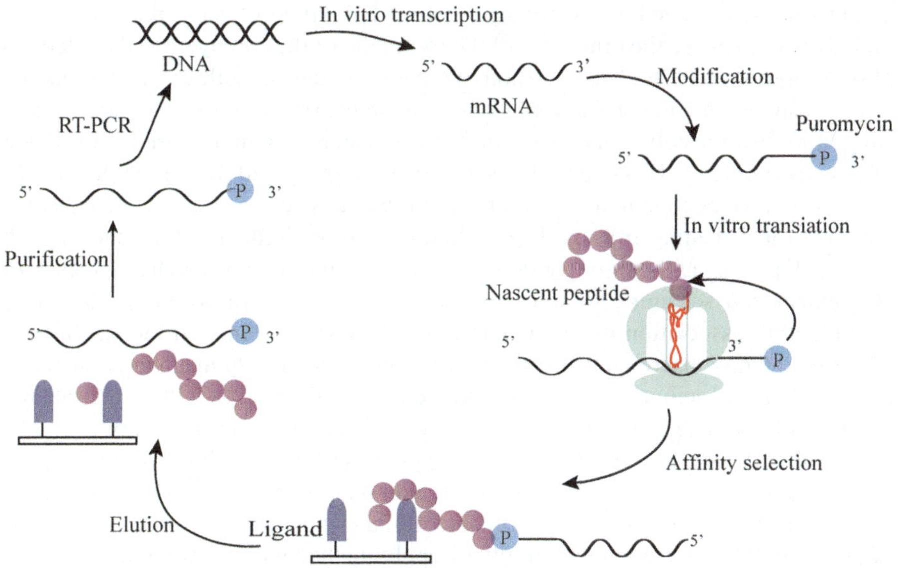

Fig. 5 Schematic representation of the mRNA-display selection round. In mRNA-display approach, the resulting mRNA is covalently conjugated with a DNA linker which contains a puromycin at its 3′-end. The generated mRNA-peptide complex is then selected by affinity chromatography and reverse-transcribed to generate DNA

between the mRNA and DNA linker. The introduced puromycin acts as a molecular analog of tyrosyl tRNA, which enters the A site of the ribosome and bonds the nascent peptide via a stable amide linkage. The resulting mRNA-protein fusion complex is directly purified by affinity chromatography and then reverse-transcribed to generate a cDNA/mRNA-protein fusion for selecting the targeted binders. Finally, the obtained DNA is used in the next cycle or sequenced for analysis.

Considering the whole selection process, several apparent advantages are evident compared with ribosome display and other related selection methods. The first point is that the constructed library can have an enormous complexity with as many as 10^{12}–10^{14} unique sequences, exceeding the complexity that can be reached by phage display. Secondly, the mRNA-peptide fusion is stable and the targeted peptides are not influenced by the complex interaction with the large ribosome. Moreover, the selection of large proteins via mRNA display provides alternative possibilities through various improved approaches. Of course, mRNA-display–based selection requires strictly a completely RNase-free environment and skillful manipulation. In addition, there is a concern that the functional ability and the enrichment efficiency of displayed proteins can be influenced by related factors, such as the charged states of the fused molecules, the initial abundance of the mRNA library, and the expression level of target peptides.

Platforms for performing mRNA display are widely established in prokaryotic and eukaryotic cell-free system as well as the PURE system. Among the typical representatives are *E. coli* extracts, wheat germ extract[126, 127], and the rabbit reticulocyte lysate system [122]. Even though the PURE system with no nucleases and proteinases showed a low efficiency of mRNA-protein fusion, the system was optimized to increase the efficiency of the formation of the conjugate in the selection of scFv antibodies [128]. The insertion of random sequences at the 5′- and 3′-coding regions can increase the efficiency by approximately 20% for mRNA-protein fusion, suggesting that the conformation of mRNA plays an important role in the formation of conjugates [129]. To increase the selection stringency and accelerate the enrichment of suitable peptide ligands, the charged yeast surface serves as a protein display platform for screening mRNA-display libraries [130]. Reducing the time of each cycle and increasing the throughput of each round can also improve the efficiency of enrichment and screening to some extent. The described transcription-translation coupled with association of puromycin-linker (TRAP) system can take as little as 14 h for selecting macrocyclic peptides with nanomolar affinity in a one-pot reaction system [131]. Moreover, a novel streptavidin-binding peptide with a non-natural amino acid was selected from the constructed non-natural peptide library using the four-base codon-mediated mRNA-display system [132], providing fruitful insights that expanded our basic understanding of non-natural peptides. Furthermore, the mRNA-display system coupled with microfluidic chip technology can yield the targeted binder with nanomolar affinity by applying the highest continuous magnetic flow for the separation[133, 134]. Therefore, there will be more updated versions based on the unique advantages of mRNA display for accelerating the exploration of therapeutic agents in the future. To date, mRNA-based display systems have been widely used in polypeptide or protein evolution, mainly including cell-penetrating peptides [135], single-chain antibodies [136], and peptide-based therapeutics [137].

3.3 cDNA Display

The cDNA display method is a modification of mRNA-display technology, and as such was seamlessly applied to screen functional binders among disulfide-rich peptides produced in a cell-free translation system [138]. The remarkable feature that differs from mRNA display is the transition from unstable mRNA-protein fusion to stable mRNA/cDNA-protein fusion through the introduction of the puromycin-linker DNA. The preparation of the puromycin-linker DNA construct mainly includes the following four parts: (1) a 5′-terminal ligation site for mRNA with T4 RNA ligase; (2) a restriction site for the release and purification of the mRNA–protein complex from the solid-phase surface; (3) a biotin moiety immobilized on a streptavidin-coated surface; (4) a 3′-terminal primer region for reverse transcription. Furthermore, two parts in this DNA-linker need to be addressed, including puromycin used for covalent linking of the nascent proteins to mRNA, and FITC moieties used for the subsequent detection and quantification.

The following section describes a selection cycle of cDNA display for in vitro selection of targeted protein binders as shown in Fig. 6 [139]. The constructed random libraries are transcribed into mRNA, which is ligated with the puromycin-linker DNA. The formed mRNA-linker fusion is translation in the cell-free translation system, yielding the mRNA-linker-protein fusion immobilized on the SA-coated solid-phase surface. Next, the resulting cDNA is produced by reverse transcription of the immobilized mRNA in the primer region of the puromycin-linker DNA. The cDNA display molecule is released from its solid surface by cleavage with the corresponding enzyme and purified via the affinity tag. In the whole selection process, the optimal design of the introduced puromycin-linker DNA plays a vital role in increasing the display efficiency and stabilizing the genotype–phenotype linkage. One pot preparation of cDNA display, designed with an RNase T1 cleavage site instead of the PvuII restriction site, improved the ligation efficiency of the mRNA-DNA fusion and facilitated the easy recovery of the cDNA-protein complexes [140]. However, there are still some limitations in the discovery of RNA-binding peptides due to contamination with RNase T1. To resolve this difficulty, deoxyinosine (dI) was added at the cleavage sites, which is digested by an endonuclease V, significantly increased the feasibility and extended the selection process, especially for RNA [141]. Another promising in vitro cDNA display system based on a cnvK-Pu-linker, which is an ultrafast photo-cross-linker using 366 nm UV irradiation instead of an enzymatic ligation reaction, was employed for selection of the FLAG epitope and an SPR-based assay [142]. To reduce the loss of specific binders during rigorous washing, the novel method of covalently photocrosslinking cDNA display molecules with their target proteins paved the way for efficient selection of peptide aptamers [143]. Emerging strategies for the above-mentioned

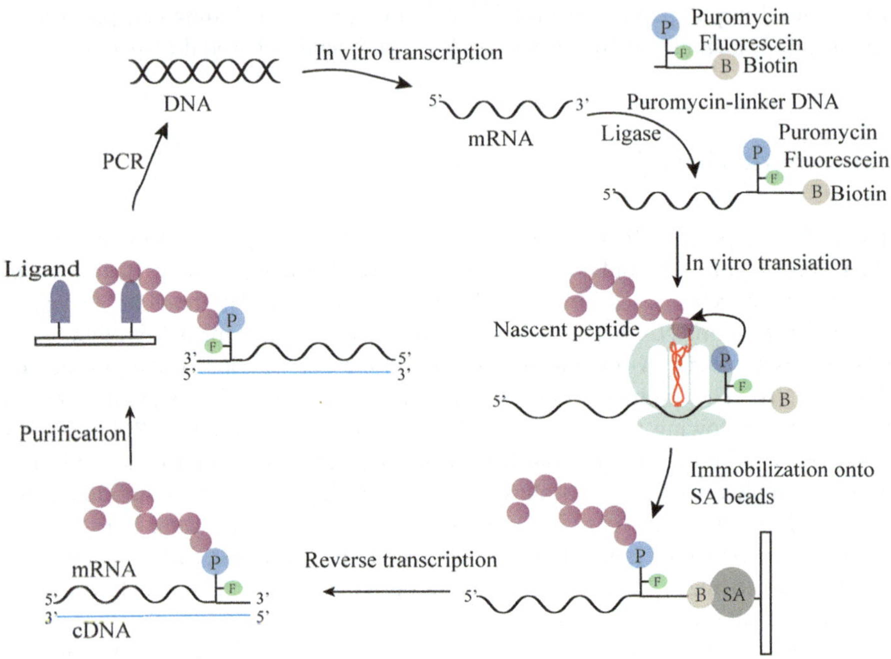

Fig. 6 Schematic representation of a cDNA display selection round. In the cDNA display approach, the constructed random libraries were transcribed into mRNA, which was ligated with the puromycin-linker DNA. The formed mRNA-linker fusion was translated in a cell-free translation system for the mRNA-linker-protein fusion. Through the binding of the SA-coated solid-phase surface, the resulting cDNA is produced by reverse transcription and released using a restriction enzyme. After the purification of fusion, the affinity binders can be selected through the immobilized target. Finally, the resulting molecules are amplified by the PCR for the next cycle

improvement could accelerate the selection in an automated high-throughput system for directed evolution.

Although cDNA can bind the mRNA to prevent its degradation in the cDNA display system to a certain extent, considering that the recovery level of cDNA is based on the initial mRNA level in the CFPS system, it is better to select appropriate extracts with low RNase activity for the in vitro cDNA display system. It was reported that the RRL system has lower nuclease activity than the *E. coli* system and the wheat germ extract used for cell-free translation [144]. Several developed CFPS platforms were adapted for in vitro cDNA display technology, such as a newly developed method of cDNA display mediated immune-PCR (cD-IPCR), which was used for detecting the model target proteins in an RRL system [145], the selection of specific single-domain antibodies against toxic gliadin using rice cell extracts [146], as well as the enrichment of the FLAG epitope motif integrated with the commercially available PUREfrex 1.0 [147]. The cDNA display approach was employed in the directed evolution of protein libraries, such as a three-finger neurotoxin (3F) protein [148], disulfide-rich peptides [138], and single-domain antibodies

[149]. The latest development of the TRAP display demonstrated the feasibility of the rapid selection of antibody-like proteins (ALPs) in various environments, even in the presence of RNases, such as RNase H [150]. At the same time, the related parameter for the 3′ end sequence, the SPC spacer length, the backbone structure in the an21 of the ALP–puromycin linker (PuL) were optimized to increase the stability of the mRNA/PuL duplex, providing an effective stratagem for facilitating the production of new ALPs.

3.4 CIS Display

The CIS display system is a DNA-based in vitro selection technology compatible with highly diverse constructed libraries [151]. The remarkable characteristic of CIS display is that the constructed DNA template can directly be linked to the nascent peptides, which offers powerful simplicity and flexibility in the selection process. The key principle of this technique takes advantage of the cis-activity of a DNA replication initiator protein (RepA), which can bind exclusively to its own DNA template. The general linear template contains the following five functional regions: promoter sequence, the selected random library, the RepA gene, the CIS element as well as the *ori* sequence (Fig. 7a). The linear DNA template in the peptide libraries can be generated by attaching a random sequence to the 5′ region of RepA. The purpose of designing the N-terminal library region ensures the intrinsic cis-activity of RepA for establishing a physical linkage of the fusion peptide and its encoding DNA when the RNA polymerase reaches the CIS element. Of course, the possibility of the designed C-terminal library has not been confirmed and further explorations are needed. The selection cycle of enriching the targets for one CIS display system is roughly divided into the following four steps (Fig. 7b). The designed DNA library is constructed with high quality and introduced into the CFPS system to generate the DNA-peptide library. For the selection process, the fusion library is incubated with the target immobilized on a solid surface. Consequently, the non-binders and weak binders are washed away for the enrichment of high-affinity binders. In the end, the eluted DNA-protein fusion complex is amplified to obtain a DNA library in the next cycle.

More importantly, this DNA-based CIS display system has three advantages over RNA-based selection. Firstly, the whole selection process does not require a strictly RNase-free environment to avoid degradation while displaying close to 10^{13} variant polypeptides. Secondly, the separation of transcription and translation steps from the purification of the protein-DNA complexes can reduce the cost and operation time. In addition, DNA display is not limited to specific cell-free platforms, such as the common *E. coli* extract [152]. The developed CIS display system has successfully achieved the selection of antibody targets [151], a high-affinity binder of VEGFR-2 [152], and other candidate therapeutic peptides [153]. Similar to the CIS display system, the proposed covalent antibody display (CAD) system contains the genetic fusions of P2A with single-chain antibodies [154]. Furthermore, CIS display

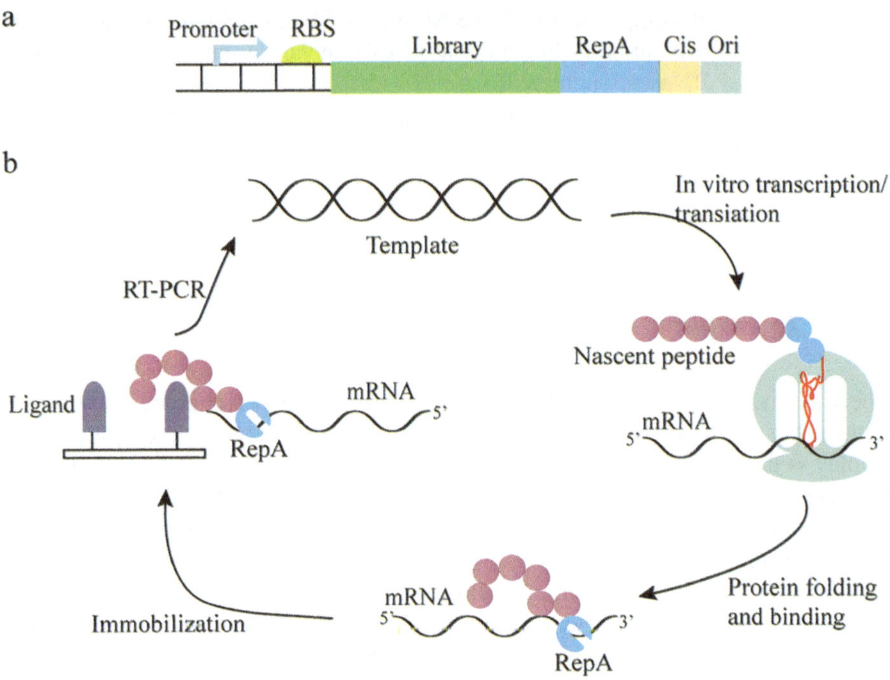

Fig. 7 Schematic representation of a CIS display selection round. (**a**) The target gene cassette was used for CIS display. The targeted template contains the following five functional regions: promoter sequence, the selected random library, the RepA gene, the CIS element as well as the ori sequence. (**b**) CIS display cycle. The designed DNA library is translated in CFPS, forming the peptide-RepA fusions. The nascent peptides can be coupled with the corresponding dsDNA template because of the binding between the RepA protein and the CIS element. The selected molecule can be captured through an immobilized target and amplified with PCR for the next round of selection

integrated with high-throughput sequencing will provide a powerful platform for the use of automation in functional selection of biomolecules [155].

In general, the four in vitro display systems mentioned above are suitable for high-throughput screening of biomolecules. These systems are characterized by the fusion of biomolecules (ribosome, mRNA, cDNA, dsDNA) with nascent proteins to form complexes that link genotypes and phenotypes for direct screening and subsequent enrichment. The main characteristics, advantages, and disadvantages of these systems have been summarized in (Table 2). In mRNA or DNA molecule-based display systems, there is a promising trend toward minimizing mRNA degradation by using DNA as bridging molecules to improve library diversity and display efficiency.

Table 2 A comparison of typical features of each cell-free display method

Approach	Ribosome display	mRNA display	cDNA display	CIS display
Library diversity	~10^{13}	10^{12}–10^{14}	>10^{12}	>10^{12}
Fusion forms	Protein–ribosome–mRNA	mRNA-protein	mRNA/cDNA-protein	dsDNA-protein
Connection type	Noncovalent	Covalent	Covalent	Covalent
Temperature range	~4°C	0–100°C	Room temperature	Room temperature
Buffer condition	High Mg^{2+}	Requires RNase-free conditions	Requires RNase-free conditions	Generally tolerant
Cell-free platform	RRL system, PURE	*E. coli* extracts, wheat germ, RRL system, PURE	Rice extracts, RRL system, PUREfrex 1.0	*E. coli* extracts,
Protein type	Antibodies and other binding proteins	Cell-penetrating peptides, single-chain antibodies, peptide-based therapeutics	Disulfide-rich peptides,	Antibodies, therapeutic peptides
Main references	[107, 109, 115]	[122, 123, 128]	[138, 139, 141]	[151, 152]

4 Conclusions and Outlook

To construct a CFPS system, platforms from different hosts, supplemented with different energy sources as well as templates with different modifications have been used in different studies, as reviewed in detail above. At the same time, the constructed CFPS platforms are classified into three categories based on the source species, including microbes, plants, and animals, which exhibited different levels of protein expression. Strategies used to stabilize the LETs and improve the expression level, such as the of engineering the genome, nuclease inhibitors, and small organic molecules were mainly summarized in this section. These adopted approaches might offer alternatives for improving the library diversity for in vitro selection. Along with the advances in vitro selection methods of the ribosome display, mRNA display, cDNA display, and CIS display, the combination of a cell-free system and high-throughput screening offers powerful tools for the selection of peptide and proteins with the desired characteristic, which was summarized and discussed in this chapter. By emphasizing these crucial elements for constructing the CFPS system and development of selection methods, this review is intended to illustrate the current state-of-the-art and point out prospective challenges in the screening of biomolecules.

Looking forward, we aim to underline the novel strategies for enhancing the protein yields that expand the application of cell-free expression systems for the high-throughput screening of biomolecules. Studies reported the in vitro evolution of protein molecules based on cell-free systems, mainly including *E. coli* extracts, RRL, wheat germ, and PURE. Aiming at higher yields and simplified preparation, *V. natriegens* extracts and the LTE system developed in recent years may be a promising platform for advancing in vitro screening [81, 101, 156]. Additionally, the emerging "TraMOS PURE" or "One-Pot PURE system" might be a viable approach to replace the commercial PURE kit for reducing costs[12, 14]. In high-throughput molecular evolution, it is worth trying to replace PEP with several low-cost and high-energy regenerative systems, such as the Cytomim system [87] and the 3-PGA system [88]. In future work, the quality of library construction and modification of linearized fragments should be emphasized to provide favorable conditions for improving the display efficiency of in vitro screening. The emphasized effective strategies detailed in Sect. 2.3 can be considered and integrated into the evolutionary screening process of protein molecules. Overall, several improvements of in vitro screening technology combined with the maturation of microarray technology will boost future progress in the process of protein evolution [157, 158]. Along with the progress of high-throughput sequencing and bioinformatic analysis tools [159], emerging innovative technologies of in vitro display methods will expand the field of biomolecular evolution and drug discovery.

Acknowledgments This work was supported by the National Key R&D Program of China (Grant No. 2019YFA0904103).

Author Contributions JJL and YHY contributed equally to this work. JJL designed the framework and drafted the original manuscript. YHY collected the related literature and drew the figs. HQ conceived the presented idea and supervised the writing of the article. All authors contributed to the article and approved the submitted version.

References

1. Galeffi P, Lombardi A, Pietraforte I, Novelli F, Di Donato M, Sperandei M, Tornambe A, Fraio-li R, Martayan A, Natali PG, Benevolo M, Mottolese M, Ylera F, Cantale C, Giacomini P (2006) Functional expression of a single-chain antibody to ErbB-2 in plants and cell-free systems. J Transl Med 4:39. https://doi.org/10.1186/1479-5876-4-3914
2. Jiang X, Ookubo Y, Fujii I, Nakano H, Yamane T (2002) Expression of Fab fragment of catalytic antibody 6D9 in an Escherichia coli in vitro coupled transcription/translation system. FEBS Lett 514(2–3):290–294. https://doi.org/10.1016/s0014-5793(02)02383-915
3. Dondapati SK, Kreir M, Quast RB, Wustenhagen DA, Bruggemann A, Fertig N, Kubick S (2014) Membrane assembly of the functional KcsA potassium channel in a vesicle-based eukaryotic cell-free translation system. Biosens Bioelectron 59:174–183. https://doi.org/10.1016/j.bios.2014.03.00418
4. Nozawa A, Nanamiya H, Miyata T, Linka N, Endo Y, Weber AP, Tozawa Y (2007) A cell-free translation and proteoliposome reconstitution system for functional analysis of plant solute transporters. Plant Cell Physiol 48(12):1815–1820. https://doi.org/10.1093/pcp/pcm15017

5. Sachse R, Dondapati SK, Fenz SF, Schmidt T, Kubick S (2014) Membrane protein synthesis in cell-free systems: from bio-mimetic systems to bio-membranes. FEBS Lett 588(17): 2774–2781. https://doi.org/10.1016/j.febslet.2014.06.00716
6. Goering AW, Li J, McClure RA, Thomson RJ, Jewett MC, Kelleher NL (2017) In vitro reconstruction of nonribosomal peptide biosynthesis directly from DNA using cell-free protein synthesis. ACS Synth Biol 6(1):39–44. https://doi.org/10.1021/acssynbio.6b0016020
7. Mikami S, Kobayashi T, Masutani M, Yokoyama S, Imataka H (2008) A human cell-derived in vitro coupled transcription/translation system optimized for production of recombinant proteins. Protein Expr Purif 62(2):190–198. https://doi.org/10.1016/j.pep.2008.09.00219
8. Jaroentomeechai T, Stark JC, Natarajan A, Glasscock CJ, Yates LE, Hsu KJ, Mrksich M, Jew-ett MC, DeLisa MP (2018) Single-pot glycoprotein biosynthesis using a cell-free tran-scription-translation system enriched with glycosylation machinery. Nat Commun 9(1):2686. https://doi.org/10.1038/s41467-018-05110-x21
9. Shimizu Y, Inoue A, Tomari Y, Suzuki T, Yokogawa T, Nishikawa K, Ueda T (2001) Cell-free translation reconstituted with purified components. Nat Biotechnol 19(8):751–755. https://doi.org/10.1038/908026
10. Shimizu Y, Kanamori T, Ueda T (2005) Protein synthesis by pure translation systems. Methods 36(3):299–304. https://doi.org/10.1016/j.ymeth.2005.04.00622
11. Wang HH, Huang PY, Xu G, Haas W, Marblestone A, Li J, Gygi SP, Forster AC, Jewett MC, Church GM (2012) Multiplexed in vivo His-tagging of enzyme pathways for in vitro single-pot multienzyme catalysis. ACS Synth Biol 1(2):43–52. https://doi.org/10.1021/sb300002926
12. Villarreal F, Contreras-Llano LE, Chavez M, Ding YF, Fan JZ, Pan TR, Tan CM (2018) Syn-thetic microbial consortia enable rapid assembly of pure translation machinery. Nat Chem Biol 14(1):29–35. https://doi.org/10.1038/Nchembio.251425
13. Shepherd TR, Du L, Liljeruhm J, Samudyata WJ, Sjodin MOD, Wetterhall M, Yomo T, Forster AC (2017) De novo design and synthesis of a 30-cistron translation-factor module. Nucleic Acids Res 45(18):10895–10905. https://doi.org/10.1093/nar/gkx75324
14. Lavickova B, Maerkl SJ (2019) A Simple, robust, and low-cost method to produce the PUR-E cell-free system. ACS Synth Biol 8(2):455–462. https://doi.org/10.1021/acssynbio.8b0042723
15. Smith GP (1985) Filamentous fusion phage: novel expression vectors that display cloned antigens on the virion surface. Science 228(4705):1315–1317. https://doi.org/10.1126/science.4001944232
16. Marks JD, Hoogenboom HR, Griffiths AD, Winter G (1992) Molecular evolution of proteins on filamentous phage. Mimicking the strategy of the immune system. J Biol Chem 267(23): 16007–16010. https://doi.org/10.1016/s0021-9258(18)41952-7230
17. Boder ET, Midelfort KS, Wittrup KD (2000) Directed evolution of antibody fragments with monovalent femtomolar antigen-binding affinity. Proc Natl Acad Sci U S A 97(20): 10701–10705. https://doi.org/10.1073/pnas.170297297233
18. Contreras-Llano LE, Tan C (2018) High-throughput screening of biomolecules using cell-free gene expression systems. Synth Biol (Oxf) 3(1):ysy012. https://doi.org/10.1093/synbio/ysy012177
19. Prevel C, Kurzawa L, Van TN, Morris MC (2014) Fluorescent biosensors for drug discovery n-ew tools for old targets – screening for inhibitors of cyclin-dependent kinases. Eur J Med Chem 88:74–88. https://doi.org/10.1016/j.ejmech.2014.10.003245
20. Doyle SK, Pop MS, Evans HL, Koehler AN (2016) Advances in discovering small molecules to probe protein function in a systems context. Curr Opin Chem Biol 30:28–36. https://doi.org/10.1016/j.cbpa.2015.10.032246
21. Longwell CK, Labanieh L, Cochran JR (2017) High-throughput screening technologies for enzyme engineering. Curr Opin Biotechnol 48:196–202. https://doi.org/10.1016/j.copbio.2017.05.012247

22. Perez JG, Stark JC, Jewett MC (2016) Cell-free synthetic biology: engineering beyond the cell. Cold Spring Harb Perspect Biol 8(12):a023853. https://doi.org/10.1101/cshperspect.a02385333
23. Nirenberg MW, Matthaei JH (1961) The dependence of cell-free protein synthesis in E. coli upon naturally occurring or synthetic polyribonucleotides. Proc Natl Acad Sci U S A 47(10): 1588–1602. https://doi.org/10.1073/pnas.47.10.158846
24. Kwon YC, Jewett MC (2015) High-throughput preparation methods of crude extract for robust cell-free protein synthesis. Sci Rep 58:663. https://doi.org/10.1038/srep0866332
25. Caschera F, Noireaux V (2014) Synthesis of 2.3 mg/ml of protein with an all Escherichia coli cell-free transcription-translation system. Biochimie 99:162–168. https://doi.org/10.1016/j.biochi.2013.11.02530
26. Guzman-Chavez F, Arce A, Adhikari A, Vadhin S, Pedroza-Garcia JA, Gandini C, Ajioka JW, Molloy J, Sanchez-Nieto S, Varner JD, Federici F, Haseloff J (2022) Constructing cell-free expression systems for low-cost access. ACS Synth Biol 11(3):1114–1128. https://doi.org/10.1021/acssynbio.1c0034231
27. Zawada JF, Yin G, Steiner AR, Yang J, Naresh A, Roy SM, Gold DS, Heinsohn HG, Murray CJ (2011) Microscale to manufacturing scale-up of cell-free cytokine production – a new appro-ach for shortening protein production development timelines. Biotechnol Bioeng 108(7):1570–1578. https://doi.org/10.1002/bit.2310335
28. Nakashima N, Tamura T (2004) Cell-free protein synthesis using cell extract of pseudomonas fluorescens and CspA promoter. Biochem Biophys Res Commun 319(2):671–676. https://doi.org/10.1016/j.bbrc.2004.05.03448
29. Wang H, Li J, Jewett MC (2018) Development of a pseudomonas putida cell-free protein synthesis platform for rapid screening of gene regulatory elements. Synth Biol (Oxf) 3(1): ysy003. https://doi.org/10.1093/synbio/ysy00349
30. Yang C, Yang M, Zhao W, Ding Y, Wang Y, Li J (2022) Establishing a Klebsiella pneumoniae-based cell-free protein synthesis system. Molecules 27(15):4684. https://doi.org/10.3390/molecules2715468450
31. Li J, Wang H, Kwon YC, Jewett MC (2017) Establishing a high yielding streptomyces-based cell-free protein synthesis system. Biotechnol Bioeng 114(6):1343–1353. https://doi.org/10.1002/bit.2625337
32. Li J, Wang H, Jewett MC (2018) Expanding the palette of Streptomyces-based cell-free protein synthesis systems with enhanced yields. Biochem Eng J 130:29–33. https://doi.org/10.1016/j.bej.2017.11.01336
33. Xu H, Liu WQ, Li J (2020) Translation related factors improve the productivity of a streptomyces-based cell-free protein synthesis system. ACS Synth Biol 9(5):1221–1224. https://doi.org/10.1021/acssynbio.0c0014038
34. Xu H, Yang C, Tian X, Chen Y, Liu WQ, Li J (2022) Regulatory part engineering for high-yield protein synthesis in an all-streptomyces-based cell-free expression system. ACS Synth Biol 11(2):570–578. https://doi.org/10.1021/acssynbio.1c0058739
35. Kelwick R, Webb AJ, MacDonald JT, Freemont PS (2016) Development of a Bacillus subtilis cell-free transcription-translation system for prototyping regulatory elements. Metab Eng 38: 370–381. https://doi.org/10.1016/j.ymben.2016.09.00881
36. Suresh G, Cabezudo I, Pulicharla R, Cuprys A, Rouissi T, Brar SK (2020) Biodegradation of aflatoxin B1 with cell-free extracts of Trametes versicolor and Bacillus subtilis. Res Vet Sci 133:85–91. https://doi.org/10.1016/j.rvsc.2020.09.00983
37. Moore SJ, MacDonald JT, Wienecke S, Ishwarbhai A, Tsipa A, Aw R, Kylilis N, Bell DJ, Mc-Clymont DW, Jensen K, Polizzi KM, Biedendieck R, Freemont PS (2018) Rapid acquisition and model-based analysis of cell-free transcription-translation reactions from nonmodel bacteria. Proc Natl Acad Sci U S A 115(19):E4340–E4349. https://doi.org/10.1073/pnas.171580611585
38. Uzawa T, Yamagishi A, Oshima T (2002) Polypeptide synthesis directed by DNA as a messeng-er in cell-free polypeptide synthesis by extreme thermophile, Thermus thermophilus

HB27 and -Sulfolobus tokodaii strain 7. J Biochem 131(6):849–853. https://doi.org/10.1093/oxfordjournals.jbchem.a00317441
39. Endoh T, Kanai T, Sato YT, Liu DV, Yoshikawa K, Atomi H, Imanaka T (2006) Cell-free protein synthesis at high temperatures using the lysate of a hyperthermophile. J Biotechnol 126(2):186–195. https://doi.org/10.1016/j.jbiotec.2006.04.01042
40. Lo Gullo G, Mattossovich R, Perugino G, La Teana A, Londei P, Benelli D (2019) Optimization of an in vitro transcription/translation system based on Sulfolobus solfataricus cell lysate. Archaea 2019:9848253. https://doi.org/10.1155/2019/984825343
41. Eagon RG (1962) Pseudomonas natriegens, a marine bacterium with a generation time of less than 10 minutes. J Bacteriol 83(4):736–737. https://doi.org/10.1128/jb.83.4.736-737.196277
42. Aiyar SE, Gaal T, Gourse RL (2002) rRNA promoter activity in the fast-growing bacterium vibrio natriegens. J Bacteriol 184(5):1349–1358. https://doi.org/10.1128/JB.184.5.1349-1358.200278
43. Des Soye BJ, Davidson SR, Weinstock MT, Gibson DG, Jewett MC (2018) Establishing a high-yielding cell-free protein synthesis platform derived from vibrio natriegens. ACS Synth Biol 7(9):2245–2255. https://doi.org/10.1021/acssynbio.8b0025256
44. Wiegand DJ, Lee HH, Ostrov N, Church GM (2018) Establishing a cell-free vibrio natriegens expression system. ACS Synth Biol 7(10):2475–2479. https://doi.org/10.1021/acssynbio.8b0022258
45. Gan R, Jewett MC (2014) A combined cell-free transcription-translation system from Saccharomyces cerevisiae for rapid and robust protein synthe. Biotechnol J 9(5):641–651. https://doi.org/10.1002/biot.201300545134
46. Hodgman CE, Jewett MC (2013) Optimized extract preparation methods and reaction conditions for improved yeast cell-free protein synthesis. Biotechnol Bioeng 110(10):2643–2654. https://doi.org/10.1002/bit.24942135
47. Schoborg JA, Hodgman CE, Anderson MJ, Jewett MC (2014) Substrate replenishment and byproduct removal improve yeast cell-free protein synthesis. Biotechnol J 9(5):630–640. https://doi.org/10.1002/biot.201300383137
48. Brodiazhenko T, Johansson MJO, Takada H, Nissan T, Hauryliuk V, Murina V (2018) Elimination of ribosome inactivating factors improves the efficiency of Bacillus subtilis and Saccharomyces cerevisiae cell-free translation systems. Front Microbiol 9:3041. https://doi.org/10.3389/fmicb.2018.03041133
49. Hodgman CE, Jewett MC (2014) Characterizing IGR IRES-mediated translation initiation for use in yeast cell-free protein synthesis. N Biotechnol 31(5):499–505. https://doi.org/10.1016/j.nbt.2014.07.001140
50. Anderson MJ, Stark JC, Hodgman CE, Jewett MC (2015) Energizing eukaryotic cell-free protein synthesis with glucose metabolism. FEBS Lett 589(15):1723–1727. https://doi.org/10.1016/j.febslet.2015.05.045141
51. Rasor BJ, Yi X, Brown H, Alper HS, Jewett MC (2021) An integrated in vivo/in vitro framework to enhance cell-free biosynthesis with metabolically rewired yeast extracts. Nat Commun 12(1):5139. https://doi.org/10.1038/s41467-021-25233-y136
52. Endo Y, Sawasaki T (2006) Cell-free expression systems for eukaryotic protein production. Curr Opin Biotechnol 17(4):373–380. https://doi.org/10.1016/j.copbio.2006.06.00964
53. Fogeron ML, Lecoq L, Cole L, Harbers M, Bockmann A (2021) Easy synthesis of complex bi-omolecular assemblies: wheat germ cell-free protein expression in structural biology. Front Mol Biosci 8:639587. https://doi.org/10.3389/fmolb.2021.63958765
54. Harbers M (2014) Wheat germ systems for cell-free protein expression. FEBS Lett 588(17):2762–2773. https://doi.org/10.1016/j.febslet.2014.05.06168
55. Goshima N, Kawamura Y, Fukumoto A, Miura A, Honma R, Satoh R, Wakamatsu A, Yamamoto J, Kimura K, Nishikawa T, Andoh T, Iida Y, Ishikawa K, Ito E, Kagawa N, Kaminaga C, Kanehori K, Kawakami B, Kenmochi K, Kimura R, Kobayashi M, Kuroita T, Kuwayama H, Maruyama Y, Matsuo K, Minami K, Mitsubori M, Mori M, Morishita R, Murase A, Nishikawa A, Nishikawa S, Okamoto T, Sakagami N, Sakamoto Y, Sasaki Y,

Seki T, Sono S, Sugiyama A, Sumiya T, Takayama T, Takayama Y, Takeda H, Togashi T, Yahata K, Yamada H, Yanagisawa Y, Endo Y, Imamoto F, Kisu Y, Tanaka S, Isogai T, Imai J, Watanabe S, Nomura N (2008) Human protein factory for converting the transcriptome into an in vitro-expressed proteome. Nat Methods 5(12):1011–1017. https://doi.org/10.1038/nmeth.127367
56. Kanoi BN, Nagaoka H, Morita M, Tsuboi T, Takashima E (2021) Leveraging the wheat germ cell-free protein synthesis system to accelerate malaria vaccine development. Parasitol Int 80: 102224. https://doi.org/10.1016/j.parint.2020.10222469
57. Buntru M, Vogel S, Spiegel H, Schillberg S (2014) Tobacco BY-2 cell-free lysate: an alternative and highly-productive plant-based in vitro translation system. BMC Biotechnol 14:37. https://doi.org/10.1186/1472-6750-14-3773
58. Buntru M, Vogel S, Stoff K, Spiegel H, Schillberg S (2015) A versatile coupled cell-free transcription-translation system based on tobacco BY-2 cell lysates. Biotechnol Bioeng 112(5):867–878. https://doi.org/10.1002/bit.2550274
59. Suzuki K, Inoue H, Matsuoka S, Tero R, Hirano-Iwata A, Tozawa Y (2020) Establishment of a cell-free translation system from rice callus extracts. Biosci Biotechnol Biochem 84(10): 2028–2036. https://doi.org/10.1080/09168451.2020.177902462
60. Murota K, Hagiwara-Komoda Y, Komoda K, Onouchi H, Ishikawa M, Naito S (2011) Arabidopsis cell-free extract, ACE, a new in vitro translation system derived from Arabidopsis callus cultures. Plant Cell Physiol 52(8):1443–1453. https://doi.org/10.1093/pcp/pcr08080
61. Cordoba-Canero D, Roldan-Arjona T, Ariza RR (2012) Using Arabidopsis cell extracts to monitor repair of DNA base damage in vitro. Methods Mol Biol 920:263–277. https://doi.org/10.1007/978-1-61779-998-3_18103
62. Liu K, Hu J (2018) Host-regulated hepatitis B virus capsid assembly in a mammalian cell-free system. Bio Protoc 8(8):e2813. https://doi.org/10.21769/BioProtoc.2813100
63. Panthu B, Ohlmann T, Perrier J, Schlattner U, Jalinot P, Elena-Herrmann B, Rautureau GJP (2018) Cell-free protein synthesis enhancement from real-time NMR metabolite kinetics: red-irecting energy fluxes in hybrid RRL systems. ACS Synth Biol 7(1):218–226. https://doi.org/10.1021/acssynbio.7b00280101
64. Anastasina M, Terenin I, Butcher SJ, Kainov DE (2014) A technique to increase protein yield in a rabbit reticulocyte lysate translation system. Biotechniques 56(1):36–39. https://doi.org/10.2144/00011412599
65. Brodel AK, Sonnabend A, Kubick S (2014) Cell-free protein expression based on extracts from CHO cells. Biotechnol Bioeng 111(1):25–36. https://doi.org/10.1002/bit.25013101
66. Ramm F, Jack L, Kaser D, Schlosshauer JL, Zemella A, Kubick S (2022) Cell-free systems enable the production of AB5 toxins for diagnostic applications. Toxins (Basel) 14(4):233. https://doi.org/10.3390/toxins1404023389
67. Thoring L, Dondapati SK, Stech M, Wustenhagen DA, Kubick S (2017) High-yield production of "difficult-to-express" proteins in a continuous exchange cell-free system based on CHO cell lysates. Sci Rep 7(1):11710. https://doi.org/10.1038/s41598-017-12188-8103
68. Yadavalli R, Sam-Yellowe T (2015) HeLa based cell free expression systems for expression of plasmodium Rhoptry proteins. J Vis Exp (100):e52772. https://doi.org/10.3791/52772269
69. Wustenhagen DA, Lukas P, Muller C, Aubele SA, Hildebrandt JP, Kubick S (2020) Cell-free synthesis of the hirudin variant 1 of the blood-sucking leech Hirudo medicinalis. Sci Rep 10(1):19818. https://doi.org/10.1038/s41598-020-76715-w272
70. Burgenson D, Gurramkonda C, Pilli M, Ge X, Andar A, Kostov Y, Tolosa L, Rao G (2018) Rap-id recombinant protein expression in cell-free extracts from human blood. Sci Rep 8(1): 9569. https://doi.org/10.1038/s41598-018-27846-8270
71. Mikami S, Masutani M, Sonenberg N, Yokoyama S, Imataka H (2006) An efficient mammalian cell-free translation system supplemented with translation factors. Protein Expr Purif 46(2):348–357. https://doi.org/10.1016/j.pep.2005.09.021105

72. Brodel AK, Sonnabend A, Roberts LO, Stech M, Wustenhagen DA, Kubick S (2013) IRES-mediated translation of membrane proteins and glycoproteins in eukaryotic cell-free systems. PloS One 8(12):e82234. https://doi.org/10.1371/journal.pone.0082234138
73. Merk H, Rues RB, Gless C, Beyer K, Dong F, Dotsch V, Gerrits M, Bernhard F (2015) Biosynthesis of membrane dependent proteins in insect cell lysates: identification of limiting parameters for folding and processing. Biol Chem 396(9–10):1097–1107. https://doi.org/10.1515/hsz-2015-010588
74. Suzuki T, Ezure T, Ando E, Nishimura O, Utsumi T, Tsunasawa S (2010) Preparation of ubiquitin-conjugated proteins using an insect cell-free protein synthesis system. J Biotechnol 145(1):73–78. https://doi.org/10.1016/j.jbiotec.2009.10.00991
75. Quast RB, Claussnitzer I, Merk H, Kubick S, Gerrits M (2014) Synthesis and site-directed fluorescence labeling of azido proteins using eukaryotic cell-free orthogonal translation systems. Anal Biochem 451:4–9. https://doi.org/10.1016/j.ab.2014.01.013139
76. Stech M, Hust M, Schulze C, Dubel S, Kubick S (2014) Cell-free eukaryotic systems for the production, engineering, and modification of scFv antibody fragments. Eng Life Sci 14(4): 387–398. https://doi.org/10.1002/elsc.20140003690
77. Zemella A, Thoring L, Hoffmeister C, Kubick S (2015) Cell-free protein synthesis: pros and cons of prokaryotic and eukaryotic systems. Chembiochem 16(17):2420–2431. https://doi.org/10.1002/cbic.20150034093
78. Mureev S, Kovtun O, Nguyen UT, Alexandrov K (2009) Species-independent translational leaders facilitate cell-free expression. Nat Biotechnol 27(8):747–752. https://doi.org/10.1038/nbt.155698
79. Cui Z, Wu Y, Mureev S, Alexandrov K (2018) Oligonucleotide-mediated tRNA sequestration enables one-pot sense codon reassignment in vitro. Nucleic Acids Res 46(12):6387–6400. https://doi.org/10.1093/nar/gky36595
80. Kovtun O, Mureev S, Jung W, Kubala MH, Johnston W, Alexandrov K (2011) Leishmania cell-free protein expression system. Methods 55(1):58–64. https://doi.org/10.1016/j.ymeth.2011.06.00697
81. Johnston WA, Moradi SV, Alexandrov K (2019) Adaption of the Leishmania cell-free expression system to high-throughput analysis of protein interactions. Methods Mol Biol 2025:403–421. https://doi.org/10.1007/978-1-4939-9624-7_1996
82. Bhide M, Natarajan S, Hresko S, Aguilar C, Bencurova E (2014) Rapid in vitro protein synthesis pipeline: a promising tool for cost-effective protein array design. Mol Biosyst 10(6): 1236–1245. https://doi.org/10.1039/c4mb00003j94
83. Kigawa T, Yabuki T, Yoshida Y, Tsutsui M, Ito Y, Shibata T, Yokoyama S (1999) Cell-free pro-duction and stable-isotope labeling of milligram quantities of proteins. FEBS Lett 442(1): 15–19. https://doi.org/10.1016/S0014-5793(98)01620-2144
84. Kim DM, Swartz JR (2001) Regeneration of adenosine triphosphate from glycolytic interme diates for cell-free protein synthesis. Biotechnol Bioeng 74(4):309-316. 129
85. Jewett MC, Swartz JR (2004) Mimicking the Escherichia coli cytoplasmic environment activates long-lived and efficient cell-free protein synthesis. Biotechnol Bioeng 86(1):19–26. https://doi.org/10.1002/bit.20026126
86. Cai Q, Hanson JA, Steiner AR, Tran C, Masikat MR, Chen R, Zawada JF, Sato AK, Hallam TJ, Yin G (2015) A simplified and robust protocol for immunoglobulin expression in Escherichia coli cell-free protein synthesis systems. Biotechnol Prog 31(3):823–831. https://doi.org/10.1002/btpr.2082140
87. Schoborg JA, Hershewe JM, Stark JC, Kightlinger W, Kath JE, Jaroentomeechai T, Natarajan A, DeLisa MP, Jewett MC (2018) A cell-free platform for rapid synthesis and testing of active oligosaccharyltransferases. Biotechnol Bioeng 115(3):739–750. https://doi.org/10.1002/bit.26502141
88. Sitaraman K, Esposito D, Klarmann G, Le Grice SF, Hartley JL, Chatterjee DK (2004) A novel cell-free protein synthesis system. J Biotechnol 110(3):257–263. https://doi.org/10.1016/j.jbiotec.2004.02.014131

89. Kim TW, Keum JW, Oh IS, Choi CY, Kim HC, Kim DM (2007) An economical and highly productive cell-free protein synthesis system utilizing fructose-1,6-bisphosphate as an energy source. J Biotechnol 130(4):389–393. https://doi.org/10.1016/j.jbiotec.2007.05.002142
90. Kim H-C, Kim T-W, Kim D-M (2011) Prolonged production of proteins in a cell-free protein synthesis system using polymeric carbohydrates as an energy source. Process Biochem 46(6): 1366–1369. https://doi.org/10.1016/j.procbio.2011.03.008139
91. Jewett MC, Calhoun KA, Voloshin A, Wuu JJ, Swartz JR (2008) An integrated cell-free metabolic platform for protein production and synthetic biology. Mol Syst Biol 4:220. https://doi.org/10.1038/msb.2008.57145
92. Akanuma G, Kobayashi A, Suzuki S, Kawamura F, Shiwa Y, Watanabe S, Yoshikawa H, Hana-i R, Ishizuka M (2014) Defect in the formation of 70S ribosomes caused by lack of ribosomal protein L34 can be suppressed by magnesium. J Bacteriol 196(22):3820–3830. https://doi.org/10.1128/Jb.01896-14147
93. Garamella J, Marshall R, Rustad M, Noireaux V (2016) The all E. coli TX-TL toolbox 2.0: a platform for cell-free synthetic biology. ACS Synth Biol 5(4):344–355. https://doi.org/10.1021/acssynbio.5b00296148
94. Michel-Reydellet N, Woodrow K, Swartz J (2005) Increasing PCR fragment stability and protein yields in a cell-free system with genetically modified Escherichia coli extracts. J Mol Microbiol Biotechnol 9(1):26–34. https://doi.org/10.1159/000088143160
95. Seki E, Matsuda N, Kigawa T (2009) Multiple inhibitory factor removal from an Escherichia co-li cell extract improves cell-free protein synthesis. J Biosci Bioeng 108(1):30–35. https://doi.org/10.1016/j.jbiosc.2009.02.011161
96. Batista AC, Levrier A, Soudier P, Voyvodic PL, Achmedov T, Reif-Trauttmansdorff T, DeVis-ch A, Cohen-Gonsaud M, Faulon JL, Beisel CL, Bonnet J, Kushwaha M (2022) Differentially optimized cell-free buffer enables robust expression from unprotected linear DNA in exonuclease-deficient extracts. ACS Synth Biol 11(2):732–746. https://doi.org/10.1021/acssynbio.1c00448159
97. Marshall R, Maxwell CS, Collins SP, Beisel CL, Noireaux V (2017) Short DNA containing c-hi sites enhances DNA stability and gene expression in E. coli cell-free transcription-translation systems. Biotechnol Bioeng 114(9):2137–2141. https://doi.org/10.1002/bit.26333152
98. Spies M, Amitani I, Baskin RJ, Kowalczykowski SC (2007) RecBCD enzyme switches lead motor subunits in response to chi recognition. Cell 131(4):694–705. https://doi.org/10.1016/j.cell.2007.09.023168
99. Amundsen SK, Spicer T, Karabulut AC, Londono LM, Eberhart C, Fernandez Vega V, Bannister TD, Hodder P, Smith GR (2012) Small-molecule inhibitors of bacterial AddAB and RecBCD helicase-nuclease DNA repair enzymes. ACS Chem Biol 7(5):879–891. https://doi.org/10.1021/cb300018x169
100. Yim SS, Johns NI, Noireaux V, Wang HH (2020) Protecting linear DNA templates in cell- free expression systems from diverse bacteria. ACS Synth Biol 9(10):2851–2855. https://doi.org/10.1021/acssynbio.0c00277157
101. Zhu B, Gan R, Cabezas MD, Kojima T, Nicol R, Jewett MC, Nakano H (2020) Increasing cell-free gene expression yields from linear templates in Escherichia coli and vibrio natriegens extracts by using DNA-binding proteins. Biotechnol Bioeng 117(12):3849–3857. https://doi.org/10.1002/bit.27538158
102. Sun ZZ, Yeung E, Hayes CA, Noireaux V, Murray RM (2014) Linear DNA for rapid prototyping of synthetic biological circuits in an Escherichia coli based TX-TL cell-free system. ACS Synth Biol 3(6):387–397. https://doi.org/10.1021/sb400131a163
103. Hong SH, Ntai I, Haimovich AD, Kelleher NL, Isaacs FJ, Jewett MC (2014) Cell-free protein synthesis from a release factor 1 deficient Escherichia coli activates efficient and multiple site-specific nonstandard amino acid incorporation. ACS Synth Biol 3(6):398–409. https://doi.org/10.1021/sb400140t162

104. Wu PS, Ozawa K, Lim SP, Vasudevan SG, Dixon NE, Otting G (2007) Cell-free transcription/translation from PCR-amplified DNA for high-throughput NMR studies. Angew Chem Int Ed Engl 46(18):3356–3358. https://doi.org/10.1002/anie.200605237170
105. Ahn JH, Chu HS, Kim TW, Oh IS, Choi CY, Hahn GH, Park CG, Kim DM (2005) Cell-free synthesis of recombinant proteins from PCR-amplified genes at a comparable productivity to that of plasmid-based reactions. Biochem Biophys Res Commun 338(3):1346–1352. https://doi.org/10.1016/j.bbrc.2005.10.094165
106. Chen X, Lu Y (2021) In silico design of linear DNA for robust cell-free gene expression. Front Bioeng Biotechnol 9:670341. https://doi.org/10.3389/fbioe.2021.670341151
107. Mattheakis LC, Bhatt RR, Dower WJ (1994) An in vitro polysome display system for identifying ligands from very large peptide libraries. Proc Natl Acad Sci U S A 91(19):9022–9026. https://doi.org/10.1073/pnas.91.19.9022215
108. Pluckthun A, Schaffitzel C, Hanes J, Jermutus L (2001) In vitro selection and evolution of proteins. Adv Protein Chem 55:367–403. https://doi.org/10.1016/s0065-3233(01)55009-3
109. Zahnd C, Amstutz P, Pluckthun A (2007) Ribosome display: selecting and evolving proteins in vitro that specifically bind to a target. Nat Methods 4(3):269–279. https://doi.org/10.1038/nmeth1003183
110. Ahangarzadeh S, Bandehpour M, Kazemi B (2017) Selection of single-chain variable fragments specific for mycobacterium tuberculosis ESAT-6 antigen using ribosome display. Iran J Basic Med Sci 20(3):327–333. https://doi.org/10.22038/ijbms.2017.8363222
111. Gan R, Jewett MC (2016) Evolution of translation initiation sequences using in vitro yeast ribosome display. Biotechnol Bioeng 113(8):1777–1786. https://doi.org/10.1002/bit.25933223
112. He M, Taussig MJ (1997) Antibody-ribosome-mRNA (ARM) complexes as efficient selection particles for in vitro display and evolution of antibody combining sites. Nucleic Acids Res 25(24):5132–5134. https://doi.org/10.1093/nar/25.24.5132221
113. Villemagne D, Jackson R, Douthwaite JA (2006) Highly efficient ribosome display selection by use of purified components for in vitro translation. J Immunol Methods 313(1–2):140–148. https://doi.org/10.1016/j.jim.2006.04.001184
114. Ohashi H, Kanamori T, Osada E, Akbar BK, Ueda T (2012) Peptide screening using PURE ribosome display. Methods Mol Biol 805:251–259. https://doi.org/10.1007/978-1-61779-379-0_14217
115. Ueda T, Kanamori T, Ohashi H (2010) Ribosome display with the PURE technology. Methods Mol Biol 607:219–225. https://doi.org/10.1007/978-1-60327-331-2_18218
116. Kanamori T, Fujino Y, Ueda T (2014) PURE ribosome display and its application in antibody technology. Biochim Biophys Acta 1844(11):1925–1932. https://doi.org/10.1016/j.bbapap.2014.04.007219
117. Luginbuhl B, Kanyo Z, Jones RM, Fletterick RJ, Prusiner SB, Cohen FE, Williamson RA, B-urton DR, Pluckthun A (2006) Directed evolution of an anti-prion protein scFv fragment to an affinity of 1 pM and its structural interpretation. J Mol Biol 363(1):75–97. https://doi.org/10.1016/j.jmb.2006.07.027225
118. Dreier B, Pluckthun A (2012) Rapid selection of high-affinity binders using ribosome display. Methods Mol Biol 805:261–286. https://doi.org/10.1007/978-1-61779-379-0_15224
119. Kunamneni A, Ogaugwu C, Bradfute S, Durvasula R (2020) Ribosome display technology: applications in disease diagnosis and control. Antibodies (Basel) 9(3):28. https://doi.org/10.3390/antib9030028185
120. Stefan N, Martin-Killias P, Wyss-Stoeckle S, Honegger A, Zangemeister-Wittke U, Pluckthun A (2011) DARPins recognizing the tumor-associated antigen EpCAM selected by phage and ribosome display and engineered for multivalency. J Mol Biol 413(4):826–843. https://doi.org/10.1016/j.jmb.2011.09.016226
121. Hammerling MJ, Fritz BR, Yoesep DJ, Kim DS, Carlson ED, Jewett MC (2020) In vitro ribosome synthesis and evolution through ribosome display. Nat Commun 11(1):1108. https://doi.org/10.1038/s41467-020-14705-2174

122. Roberts RW, Szostak JW (1997) RNA-peptide fusions for the in vitro selection of peptides and proteins. Proc Natl Acad Sci U S A 94(23):12297–12302. https://doi.org/10.1073/pnas.94.23.12297228
123. Wang R, Cotten SW, Liu R (2012) mRNA display using covalent coupling of mRNA to translated proteins. Methods Mol Biol 805:87–100. https://doi.org/10.1007/978-1-61779-379-0_6192
124. Kurz M, Gu K, Lohse PA (2000) Psoralen photo-crosslinked mRNA–puromycin conjugates: a novel template for the rapid and facile preparation of mRNA–protein fusions. Nucleic Acids Res 28(18):e83–e83. https://doi.org/10.1093/nar/28.18.e83237
125. Moore MJ, Sharp PA (1992) Site-specific modification of pre-mRNA: the 2′-hydroxyl groups at the splice sites. Science 256(5059):992–997. https://doi.org/10.1126/science.1589782236
126. Miyamoto-Sato E, Takashima H, Fuse S, Sue K, Ishizaka M, Tateyama S, Horisawa K, Sawasaki T, Endo Y, Yanagawa H (2003) Highly stable and efficient mRNA templates for mRNA-protein fusions and C-terminally labeled proteins. Nucleic Acids Res 31(15):e78. https://doi.org/10.1093/nar/gng078234
127. Seelig B (2011) mRNA display for the selection and evolution of enzymes from in vitro-translated protein libraries. Nat Protoc 6(4):540–552. https://doi.org/10.1038/nprot.2011.312235
128. Nagumo Y, Fujiwara K, Horisawa K, Yanagawa H, Doi N (2016) PURE mRNA display for in vitro selection of single-chain antibodies. J Biochem 159(5):519–526. https://doi.org/10.1093/jb/mvv131198
129. Takahashi K, Sunohara M, Terai T, Kumachi S, Nemoto N (2017) Enhanced mRNA-protein fusion efficiency of a single-domain antibody by selection of mRNA display with additional rand-om sequences in the terminal translated regions. Biophys Physicobiol 14:23–28. https://doi.org/10.2142/biophysico.14.0_23199
130. Bacon K, Bowen J, Reese H, Rao BM, Menegatti S (2020) Use of target-displaying magnetized yeast in screening mRNA-display peptide libraries to identify ligands. ACS Comb Sci 22(12):738–744. https://doi.org/10.1021/acscombsci.0c00171201
131. Ishizawa T, Kawakami T, Reid PC, Murakami H (2013) TRAP display: a high-speed selection method for the generation of functional polypeptides. J Am Chem Soc 135(14):5433–5440. https://doi.org/10.1021/ja312579u238
132. Muranaka N, Hohsaka T, Sisido M (2006) Four-base codon mediated mRNA display to construct peptide libraries that contain multiple nonnatural amino acids. Nucleic Acids Res 34(1):e7. https://doi.org/10.1093/nar/gnj003202
133. Tabata N, Sakuma Y, Honda Y, Doi N, Takashima H, Miyamoto-Sato E, Yanagawa H (2009) Rapid antibody selection by mRNA display on a microfluidic chip. Nucleic Acids Res 37(8): e64-e64. https://doi.org/10.1093/nar/gkp184241
134. Olson CA, Nie J, Diep J, Al-Shyoukh I, Takahashi TT, Al-Mawsawi LQ, Bolin JM, Elwell AL, Swanson S, Stewart R, Thomson JA, Soh HT, Roberts RW, Sun R (2012) Single-round, multiplexed antibody mimetic design through mRNA display. Angew Chem Int Ed 51(50): 12449–12453. https://doi.org/10.1002/anie.201207005240
135. Kamide (2009) Isolation of novel cell-penetrating peptides from a random peptide library using in vitro virus and their modifications. Int J Mol Med 25(01):41–51. https://doi.org/10.3892/ijmm_00000311243
136. Fukuda I, Kojoh K, Tabata N, Doi N, Takashima H, Miyamoto-Sato E, Yanagawa H (2006) In vitro evolution of single-chain antibodies using mRNA display. Nucleic Acids Res 34(19): e127. https://doi.org/10.1093/nar/gkl618242
137. Duan H, Kachko A, Zhong L, Struble E, Pandey S, Yan H, Harman C, Virata-Theimer ML, De-ng L, Zhao Z, Major M, Feinstone S, Zhang P (2012) Amino acid residue-specific neutralization and nonneutralization of hepatitis C virus by monoclonal antibodies to the E2 protein. J Virol 86(23):12686–12694. https://doi.org/10.1128/JVI.00994-12244
138. Yamaguchi J, Naimuddin M, Biyani M, Sasaki T, Machida M, Kubo T, Funatsu T, Husimi Y, Nemoto N (2009) cDNA display: a novel screening method for functional disulfide-rich

peptides by solid-phase synthesis and stabilization of mRNA-protein fusions. Nucleic Acids Res 37(16):e108. https://doi.org/10.1093/nar/gkp514203
139. Mochizuki Y, Nemoto N (2012) Evolution of disulfide-rich peptide aptamers using cDNA display. Methods Mol Biol 805:237–250. https://doi.org/10.1007/978-1-61779-379-0_13211
140. Mochizuki Y, Biyani M, Tsuji-Ueno S, Suzuki M, Nishigaki K, Husimi Y, Nemoto N (2011) One-pot preparation of mRNA/cDNA display by a novel and versatile puromycin-linker DNA. ACS Comb Sci 13(5):478–485. https://doi.org/10.1021/co2000295214
141. Ueno S, Kimura S, Ichiki T, Nemoto N (2012) Improvement of a puromycin-linker to extend the selection target varieties in cDNA display method. J Biotechnol 162(2–3):299–302. https://doi.org/10.1016/j.jbiotec.2012.09.003204
142. Mochizuki Y, Suzuki T, Fujimoto K, Nemoto N (2015) A versatile puromycin-linker using cnvK for high-throughput in vitro selection by cDNA display. J Biotechnol 212:174–180. https://doi.org/10.1016/j.jbiotec.2015.08.020206
143. Terai T, Koike T, Nemoto N (2020) Photocrosslinking of cDNA display molecules with their target proteins as a new strategy for peptide selection. Molecules 25(6):1472. https://doi.org/10.3390/molecules25061472209
144. Hino M, Kataoka M, Kajimoto K, Yamamoto T, Kido J, Shinohara Y, Baba Y (2008) Efficiency of cell-free protein synthesis based on a crude cell extract from Escherichia coli, wheat germ, and rabbit reticulocytes. J Biotechnol 133(2):183–189. https://doi.org/10.1016/j.jbiotec.2007.08.008250
145. Anzai H, Terai T, Jayathilake C, Suzuki T, Nemoto N (2019) A novel immuno-PCR method using cDNA display. Anal Biochem 578:1–6. https://doi.org/10.1016/j.ab.2019.04.017252
146. Jayathilake C, Kumachi S, Arai H, Motohashi M, Terai T, Murakami A, Nemoto N (2020) In vitro selection of anti-gliadin single-domain antibodies from a naive library for cDNA-display mediated immuno-PCR. Anal Biochem 589:113490. https://doi.org/10.1016/j.ab.2019.113490253
147. Reyes SG, Kuruma Y, Fujimi M, Yamazaki M, Eto S, Nishikawa S, Tamaki S, Kobayashi A, Mizuuchi R, Rothschild L, Ditzler M, Fujishima K (2021) PURE mRNA display and cDNA dis-play provide rapid detection of core epitope motif via high-throughput sequencing. Biotechnol Bioeng 118(4):1736–1749. https://doi.org/10.1002/bit.27696251
148. Naimuddin M, Kobayashi S, Tsutsui C, Machida M, Nemoto N, Sakai T, Kubo T (2011) Direct-ed evolution of a three-finger neurotoxin by using cDNA display yields antagonists as well as agonists of interleukin-6 receptor signaling. Mol Brain 4(1):2. https://doi.org/10.1186/1756-6606-4-2249
149. Nemoto N, Kumachi S, Arai H (2018) In vitro selection of single-domain antibody (VHH) using cDNA display. Methods Mol Biol 1827:269–285. https://doi.org/10.1007/978-1-4939-8648-4_14210
150. Kondo T, Eguchi M, Kito S, Fujino T, Hayashi G, Murakami H (2021) cDNA TRAP display for rapid and stable in vitro selection of antibody-like proteins. Chem Commun (Camb) 57(19):2416–2419. https://doi.org/10.1039/d0cc07541h212
151. Odegrip R, Coomber D, Eldridge B, Hederer R, Kuhlman PA, Ullman C, FitzGerald K, McGre-gor D (2004) CIS display: in vitro selection of peptides from libraries of protein-DNA complexes. Proc Natl Acad Sci U S A 101(9):2806–2810. https://doi.org/10.1073/pnas.0400219101262
152. Patel S, Mathonet P, Jaulent AM, Ullman CG (2013) Selection of a high-affinity WW domain against the extracellular region of VEGF receptor isoform-2 from a combinatorial library using CIS display. Protein Eng Des Sel 26(4):307–315. https://doi.org/10.1093/protein/gzt003256
153. Mathonet P, Ioannou A, Betley J, Ullman C (2011) CIS display, a DNA-based in vitro selection technology for therapeutic peptides. Chim Oggi-Chem Today 29(2):10–12. 267. https://www.researchgate.net/publication/233734623
154. Reiersen H, Lobersli I, Loset GA, Hvattum E, Simonsen B, Stacy JE, McGregor D, Fitzgerald K, Welschof M, Brekke OH, Marvik OJ (2005) Covalent antibody display – an

in vitro antibody-DNA library selection system. Nucleic Acids Res 33(1):e10. https://doi.org/10.1093/nar/gni010259
155. Galan A, Comor L, Horvatic A, Kules J, Guillemin N, Mrljak V, Bhide M (2016) Library-based display technologies: where do we stand? Mol Biosyst 12(8):2342–2358. https://doi.org/10.1039/c6mb00219f258
156. Norouzi M, Panfilov S, Pardee K (2021) High-efficiency protection of linear DNA in cell-free extracts from Escherichia coli and vibrio natriegens. ACS Synth Biol 10(7):1615–1624. https://doi.org/10.1021/acssynbio.1c0011057
157. Fan J, Villarreal F, Weyers B, Ding Y, Tseng KH, Li J, Li B, Tan C, Pan T (2017) Multi-dimensional studies of synthetic genetic promoters enabled by microfluidic impact printing. Lab Chip 17(13):2198–2207. https://doi.org/10.1039/c7lc00382j266
158. Syu GD, Dunn J, Zhu H (2020) Developments and applications of functional protein microarrays. Mol Cell Proteomics 19(6):916–927. https://doi.org/10.1074/mcp.R120.001936265
159. Blanco C, Verbanic S, Seelig B, Chen IA (2020) High throughput sequencing of in vitro select-ions of mRNA-displayed peptides: data analysis and applications. Phys Chem Chem Phys 22(12):6492–6506. https://doi.org/10.1039/c9cp05912a193

Adv Biochem Eng Biotechnol (2023) 185: 91–128
https://doi.org/10.1007/10_2023_224

Published online: 13 June 2023

Cell-Free Production Systems in Droplet Microfluidics

Rémi Sieskind, Aitziber L. Cortajarena, and Aitor Manteca

Contents

Abstract The use of cell-free production systems in droplet microfluidic devices has gained significant interest during the last decade. Encapsulating DNA

R. Sieskind
Institut Pasteur, Université de Paris, Unité d'Architecture et de Dynamique des Macromolécules Biologiques, Paris, France

A. L. Cortajarena
Center for Cooperative Research in Biomaterials (CIC biomaGUNE), Basque Research and Technology Alliance (BRTA), Donostia-San Sebastián, Spain

IKERBASQUE, Basque Foundation for Science, Bilbao, Spain

A. Manteca (✉)
Center for Cooperative Research in Biomaterials (CIC biomaGUNE), Basque Research and Technology Alliance (BRTA), Donostia-San Sebastián, Spain
e-mail: amanteca@cicbiomagune.es

replication, RNA transcription, and protein expression systems in water-in-oil drops allows for the interrogation of unique molecules and high-throughput screening of libraries of industrial and biomedical interest. Furthermore, the use of such systems in closed compartments enables the evaluation of various properties of novel synthetic or minimal cells. In this chapter, we review the latest advances in the usage of the cell-free macromolecule production toolbox in droplets, with a special emphasis on new on-chip technologies for the amplification, transcription, expression, screening, and directed evolution of biomolecules.

Graphical Abstract

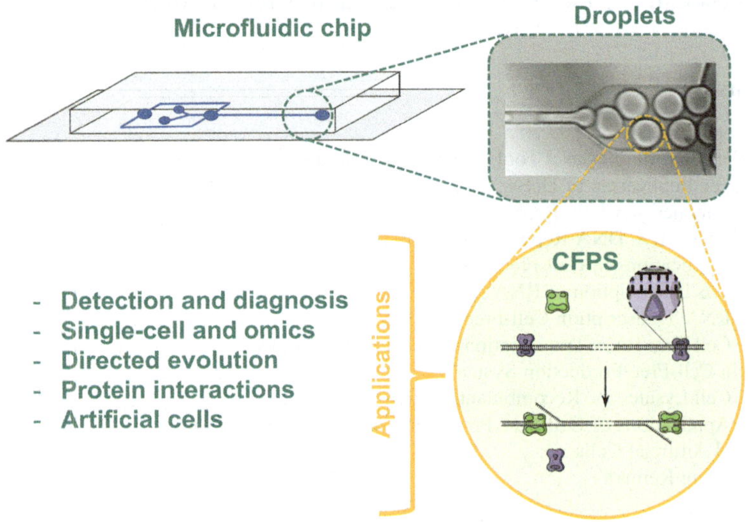

Keywords Cell-free production system, Directed evolution, DNA replication, Droplet microfluidics, Protein expression, Protein screening

1 Introduction

The usage of cell-free production systems (CFPS) has experienced exponential growth in recent years, leading to a revolution in the field of digital biology. In these systems, biomolecules such as DNA, RNA, and proteins are synthesized in vitro without the need for a host organism. Although the reactions of these systems can be easily implemented in a test tube, coupling them to mid- or high-throughput techniques results in a huge improvement in the versatility and complexity of the experiments that can be performed. Following this approach, the use of microarrays [1, 2], robotic automation processes [3, 4], and especially microfluidics has gained capital importance in the experimental design in the cell-free biology field. Furthermore, the encapsulation of many standard analytical techniques has led to a plethora of digital methods, improving the throughput of these techniques by orders of magnitude.

In a historical context, in vitro DNA amplification was the first methodology to be exploited. The development of numerous tools dedicated to the in vitro amplification of DNA in the past 40 years constitutes a revolution for many fields of the natural sciences such as molecular cloning, sequencing, and in vitro diagnostics, among others. These tools are all based on specific properties of different DNA polymerases such as processivity, thermal stability, strand displacement activity, or exonuclease performance. Despite the large diversity of mechanisms implemented over the years, all these methods have in common the ability to create, exponentially, a large number of copies of a target DNA molecule. Polymerase chain reaction (PCR) [5], based on specific primers, thermostable polymerases, and cycles of incubation at different temperatures, allows the generation of millions of copies of a user-defined segment of an initial DNA molecule. Today, the expansion of this method has led to a myriad of different techniques, with linear or circular amplicons, isothermal reactions, and non-specific processes. For instance, to overcome certain limitations such as the length of the amplicons (hardly greater than 10 kb) or the necessity of a heavy and costly thermocycler, isothermal methods have been created. These techniques based on a more complex design of primers, highly processive and strand-displacing polymerases, and sometimes complementary unwinding enzymes offer the opportunity to exponentially amplify DNA in wider variety of environments, paving the way for many novel applications.

For all these technologies, the generation of water-in-oil picoliter droplets using microfluidics represents a good opportunity to miniaturize and parallelize the reactions, opening the way to digital assays and many single-cell processes or analyses. By diluting the target molecule in an emulsion to a concentration of approximately one molecule per droplet, it is possible to amplify different DNA molecules separately before an upstream process, such as in vitro transcription and translation. This *in emulsio* amplification coupled with a fluorescent reporter also allows for the establishment of solid statistics regarding the droplet occupancy and to access to an absolute quantification of the target.

Because of its lower stability compared to its DNA counterpart, RNA inspired less technological advances in the last few decades. The study of single-cell transcriptomes and of certain RNA biomarkers, such as micro-RNA (miRNA) or circular RNA (circRNA), has been greatly developed in the last few years and remains a hot topic. Nevertheless, these studies rely on a reverse-transcription step before classical DNA handling. The recent SARS-CoV2 outbreak and the development of mRNA-based vaccines favored the emergence of a growing interest in RNA biotechnologies. Cell-free transcription revealed to be of great interest for the production of mRNA for vaccine technologies, therapeutics such as aptamers or silencing RNA (siRNA), which present the advantage of being degradable and not being integrable into the host genome. In the domain of artificial cells, the design of riboswitches, mRNA secondary structures able to up- or downregulate protein expression in response to chemical stimuli, widens the range of possible cellular functions.

In the discipline of in vitro protein expression, two types of CFPS can be distinguished for protein production depending on the way they have been produced. On one side, cell extract systems [6] profit the extract of a target organism as the source for the transcriptional/translational machinery, such as ribosomes and DNA and RNA polymerases, together with the necessary enzymes, energy systems, and cofactors to complete the transcription and translation of a DNA into a protein. In this way, a plethora of lysates and cell-free reconstituted systems from different organisms are now commercially available. On the other side, protein synthesis using recombinant elements (PURE) systems [7] produce and purify all the necessary biomolecular machinery (ribosomes, polymerases, etc.) proteins, and cofactors individually, which are then mixed together with the necessary energy systems at the required concentration.

CFPSs have been widely used in several high-throughput protein engineering applications over the last few years, such as examinations of protein libraries [8], protein–protein [9] interactions, and directed evolution of biomolecules [10]. In this sense, droplet microfluidics has become an emerging technology for high-throughput biomolecules analysis. In a droplet microfluidic system, a water-based reaction mix containing the desired biomolecules and a carrier fluid (typically fluorinated oil) is injected into a microfluidic device to form droplets. The volume of these droplets can be precisely controlled (ranging from few picoliters to several nanoliters). These drops possess exceptional stability, thanks to the fluorosurfactants [11–13] that have been previously added to the oil and can be incubated for several days. Thus, they can be used as microreactors to perform biochemical reactions. Moreover, additional reagents could be sequentially added by means of picoinjection or droplet-fusion microfluidic chips.

In this chapter, we review the latest advances combining CFPSs and droplet microfluidic methods in the fields of biochemistry and biophysics (Fig. 1). A special weight is dedicated to novel on-chip methods for the detection, production, and optimization of biologically relevant molecules.

2 Droplet Microfluidics as a Tool for Biological Studies

Droplet microfluidics refers to microelectromechanical systems able to create, mix, control, and evaluate water-in-oil droplets inside a microfluidic chip. The use of such systems has experienced remarkable progress during the last two decades due to the increase of global sequencing capacities and the boom of several omics, among other reasons. Despite the variety of these systems is huge, the microfluidic workflow is many times similar, and the processes can be classified into four groups:

1. *Compartmentalization.* The first step in every droplet microfluidics workflow is to encapsulate the biomolecules of interest. Isolating unique molecules in such

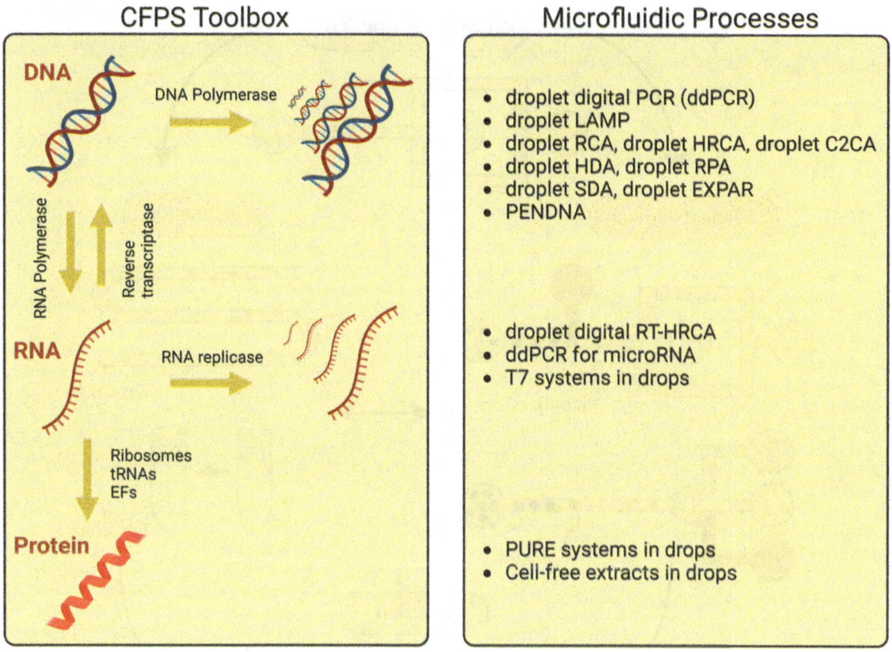

Fig. 1 Cell-free macromolecular synthesis in droplets. The left panel shows the main enzymes and molecular machines that take part in the in vitro replication, transcription, and translation processes. The right panel depicts the applications of these biomolecules in droplet microfluidics processes (Created with BioRender.com)

closed compartments will allow the user to interrogate these molecules in further processes. Depending on the following steps, some cell-free reagents can be already included in this step. These chips are commonly called drop makers. Several geometries have been used to generate drops, being the T-junction [14] and the flow-focusing [15] geometries the most used ones (Fig. 2 top).

2. *Reagent addition.* Once the biomolecules of interest have been compartmental ized, it is necessary to add certain reagents to amplify, transcribe, and/or express the initial biomolecules into the final products we want to analyze. Two of the most popular methods to add regents individually to each droplet are droplet picoinjection and droplet fusion. In the picoinjection method [16] (Fig. 2 left), a channel with the desired reagents is focused toward another perpendicular channel where a flow of previously injected drops is running. A pair of positive and negative electrodes destabilizes the water/oil interface of the drops and allows the injection of the reagents. Additional shielding electrodes can be added to avoid the coalescence of the initial drop flow. On the other hand, droplet fusion [17] (Fig. 2 left) exploits the same electric drop destabilization principle to electrocoalesce two different drops. In this method, two different flows of drops are injected and focused toward a chamber where the fusion takes place by means of a pair of electrodes. Again, extra shielding electrodes can also be added.

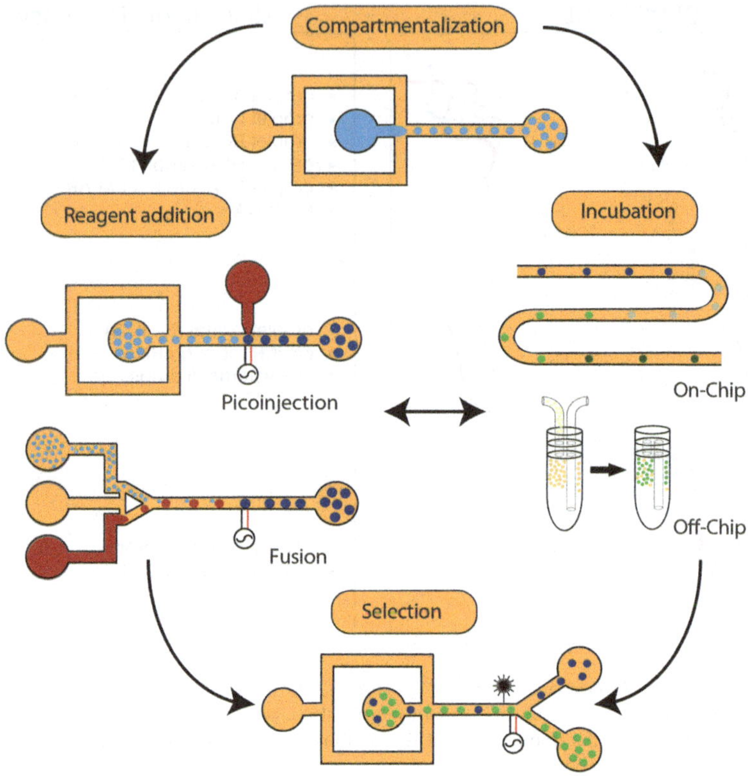

Fig. 2 Droplet microfluidics workflow for biological processes. The process starts with the encapsulation of the biomolecule of interest with the desired reagents. After this, it continues with one or several reagent additions and/or incubation steps. If the aim of the applications is to select biomolecules with improved properties, a final sorting step must be carried out

3. *Incubation.* The reagents that compose CFPS must be often subjected to one (protein expression systems) or several (PCR methods) incubation cycles. These incubations can be carried out inside a chip [18] (on-chip) or outside (off-chip) (Fig. 2 right). The steps of reagent addition and incubation are usually sequentially repeated as many times as the microfluidic workflow requires these steps.
4. *Selection.* After all the reagent additions and incubations are completed, the microfluidic workflow ends up with a screening and selection process. The most used method is the fluorescence-activated droplet sorting (FADS) [19, 20] (Fig. 2 down). Here, it is necessary to establish a fluorescence assay that generates a readout signal proportional to the fitness of the analyzed variant. Nevertheless, some considerations must be considered to implement this microfluidic method. First, a minimal activity is required to properly screen and sort the drops. Second, the speed of the readout should be fast enough to allow the high-throughput sorting of the drops by the electrodes. And third, the diffusion of the CFPS reagents and the biomolecules of interest between the droplets and the carrying

oil must be avoided [21]. Modified forms of the commercially available fluorescence-activated cell sorting (FACS) [22, 23] or benchtop-flow cytometers with water-in-oil-in-water (W/O/W) [24–26] double emulsions drops have been also used to profit the various lasers of this type of equipment.

The use of this type of workflows permits the user to work in the sub-millisecond timescale down to 1 nM of product and the kilohertz range [27]. Nevertheless, employing a fluorogenic assay is not always easy and it is strictly limited to the use of a fluorophore. This can be performed using three different methodologies: (1) with a direct measurement of a fluorogenic substrate [27–29], (2) by means of a coupled assay [30, 31], and (3) or with the liberation of a quencher [23]. Moreover, other detection systems based on different physical properties of the drops have been explored, always compatible with microfluidic chips. As a result, measurements of absorption [32], fluorescence polarization [33], light scattering [34], electrochemistry [35], mass spectroscopy [36, 37], Raman spectroscopy [38], and even image processing coupled to artificial intelligence [39–41] have been developed during the last years.

3 Cell-Free Replication of DNA

In this chapter, we will briefly describe the seminal paper of PCR in the successive development of other techniques and focus on isothermal methods in drops, compatible with posterior in vitro transcription and translation steps.

3.1 Polymerase Chain Reaction (PCR)

First published in 1986, PCR relies on multiple repetitions of denaturation, primers annealing, and elongation steps to perform exponential amplification of a specific user-defined fragment of DNA [5]. At that time, this discovery represented extraordinary progress for DNA handling and impacted widely the natural sciences with such importance that it was awarded with the Nobel Prize of Chemistry in 1993 [42]. The process consists of three well-defined steps: (1) The denaturation step consists in a high-temperature incubation to separate the two complementary molecules of DNA. (2) After this, specific primers binding at each end of the fragment of interest can then anneal to the single-stranded template and recruit a polymerase during an incubation at a lower temperature. (3) Finally, the temperature is increased to reach an optimal activity of the polymerase that complements each strand leading to multiplication by two of the DNA molecules of interest. This method is nowadays ubiquitous in molecular biology laboratories and gave birth to numerous new tools such as real-time or quantitative PCR (qPCR) [43] for relative DNA quantification (widely used in diagnostics), reverse-transcription PCR (rtPCR) [44] for the mRNA processing necessary for transcriptomics but also clinical diagnostics, or error-prone PCR (epPCR) [45] for random mutants libraries generation underlying the growing field of directed evolution.

3.2 Alternative DNA Replication Techniques

Loop-mediated isothermal amplification (LAMP) [46] constitutes an alternative to PCR to exponentially replicate a specific, user-defined amplicon at a constant temperature thanks to two pairs of primers and a polymerase exhibiting a high-strand displacement activity (Fig. 3). One pair of tail-containing primers is used to create two single-stranded DNA molecules (forward and reverse) presenting at each end a hairpin secondary structure. One of the two hairpins is self-priming while the same primers are used to initiate the replication in the opposite direction. The second pair of primers binds the template beyond the first pair and is used for initiating the whole process by displacing the tailed amplicon. All this complex process produces longer and longer concatemers of the amplicon and produces a much higher quantity of double-stranded DNA than PCR. Despite the potentially challenging primer design (sometimes facilitated by many online and proprietary tools), the simplicity of the isothermal process, the robustness, and the high productiveness of this method makes it well-suited for many point-of-care devices or field diagnostics.

Inspired by the replication processes of certain viruses and based on polymerases with a high-strand displacement activity, *rolling-circle amplification (RCA)* enables the production of a long linear single-stranded concatemeric complementary DNA molecules of a circular template [47] (Fig. 4a top). This linear process, coupled to proper fluorescent probes, has been used for diagnostics, in conjugation with solid-phase immunoassays where it played the role of signal amplifier. One way to improve this technique and create an exponential process is the use of one or several specific reverse primers that allow us to complement the long single-stranded molecule. This process, called *hyperbranched rolling-circle amplification (HRCA)* [48], makes the polymerase peel off the complementary strand synthesized at the previous turn around the circular template, producing longer and longer single-stranded complementary DNAs that can be bound by the initial primer and continuously (Fig. 4a bottom).

In the same perspective, the addition of an oligonucleotide binding the single-stranded concatemer at a restriction enzyme locus coupled to a ligase allows us to monomerize and circularize the produced DNA, creating new circular templates. This process is called *circle-2-circle amplification (C2CA)* [49]. To increase the number of priming sites, or when the sequence of the genetic material is unknown, it is possible to draw on short random primers in the HRCA process. This very productive technique, called *multiple displacement amplification (MDA)* [50], has been extensively used for whole-genome amplification and sequencing (Fig. 4a bottom). Interestingly, the use of random primers allows us to release the constraint

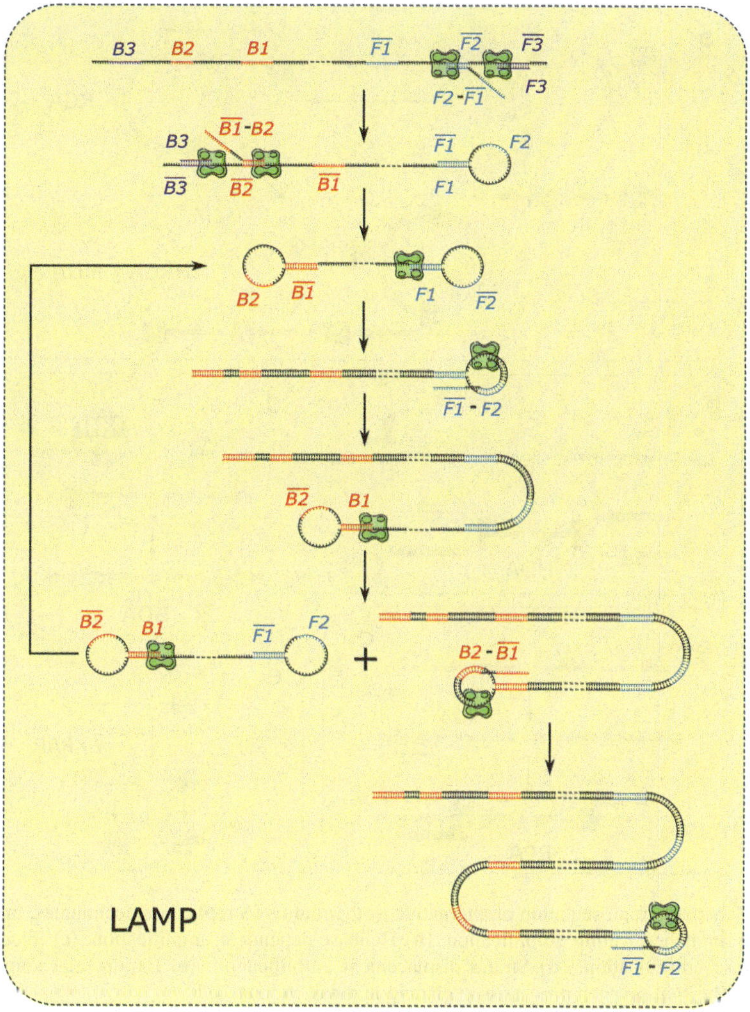

Fig. 3 Schematic representation of the loop-mediate isothermal amplification (LAMP). This process uses two pairs of primers (B1, B2 and F1, F2) and a specific polymerase with high-strand displacement activity

on the circularity of the template. This technique has been successfully used for the amplification of both circular and linear amplicons [51].

The denaturation step at the beginning of each PCR cycle allows us to unwind the DNA, giving the primers and polymerases the opportunity to invade the transiently single strands of the template. To overcome the lack of this step in isothermal process, such as in many natural replisomes, it is also possible to draw upon accessory proteins in order to unwind the DNA and stabilize the obtained single-stranded molecule. Following this idea, *helicase-dependent amplification (HDA)*

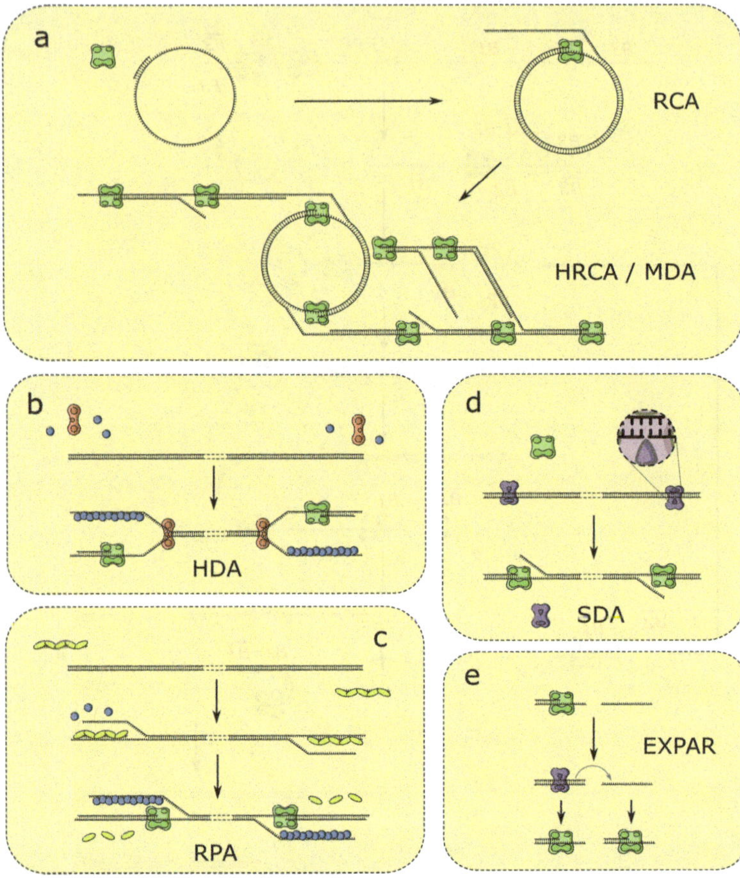

Fig. 4 Schematic representation of alternative isothermal DNA replication techniques. (**a**) Rolling and hyperbranched rolling amplification. (**b**) Helicase-dependent amplification. (**c**) Recombinase polymerase amplification. (**d**) Strand displacement amplification. (**e**) Exponential amplification reaction. The step on each process is explained in detail in Sect. 3.2

[52] was created. It uses the helicase UvrD from *Escherichia coli* to enzymatically unwind the double-stranded DNA and allows efficient primer hybridization (Fig. 4b). Contrarily to LAMP or RCA-derivatives, it works with relatively simple primers, similar to the ones used in PCR, and the product is linear and with a fixed length. Similarly, *recombinase polymerase amplification (RPA)* [53] also dispenses with thermal denaturation thanks to an additional enzyme (Fig. 4c). This technique relies on the T4 phage UvsX recombinase that binds the primers and enzymatically forces the annealing to the template. The product is also linear and the primers similar to PCR. At the closest to natural replication, several teams characterized in vitro versions of complete replisomes of phages: the pioneering work of Margarita Salas with Phi29 [54], T7 [55], or T4 [22]; prokaryotes: *Bacillus subtilis* [56], *E. coli* [57, 58]; and even eukaryotes: *Saccharomyces cerevisiae* [59]. In contrast with HDA

Table 1 Comparison between isothermal DNA amplification techniques. The temperature, the amplification rate, and the time consumed in the reaction have been evaluated

Technique	Temperature (°C)	Amplification rate	Time (min)	Reference
LAMP	65	10^9	60	[46]
RCA	30	10^9	120	[47]
HDA	37	10^6	120	[52]
RPA	37	10^9	20	[53]
SDA	37	10^7	120	[60]

and RPA, these amplification methods have not been implemented in an application-oriented manner. And yet, several of them start now being used in larger frameworks, as discussed below.

To replicate DNA at a constant temperature, the use of a polymerase presenting a high-strand displacement activity in conjugation with restriction enzymes constitutes another possibility. In 1992, early in the history of the exponential DNA amplification, Walker et al. published a method based on the ability of HincII to cleave only the unprotected strand of its hemiphosphothiolated recognition site (the apparition of natural and engineered nicking enzymes simplified the original process) and the capacity of the *E. coli* DNA polymerase I (exo-) to start from a nick and to displace a DNA strand. This method, named *strand displacement amplification (SDA)* [60], consists of using primers presenting a tail that contains the endonuclease recognition site (Fig. 4d). When the nicking enzyme cuts, the polymerase extends the 3′-end formed, displaces the strand in front of it, and releases a single-stranded copy of the target that can be complemented thanks to the reverse primer. As the process happens for both strands, the amplification is exponential and yields linear fragments.

The same principle of nicking/elongation can be used to exponentially amplify short single-stranded oligonucleotides, with important constraints on the template design. Starting from a primer ("a") and a template strand that consists in a repetition of a sequence that, once concatenated, makes the recognition sequence of the nicking enzyme appear ("/a/a"), it is possible to perform an *exponential amplification reaction (EXPAR)* [61]. When "a" binds "/a/a," it is elongated by the polymerase; the nicking enzyme cuts and releases two "a" strands (Fig. 4e). Such a reaction, called autocatalytic, can be cascaded and coupled to different modules to create complex behaviors monitored by specific fluorescent probes. This is the principle of *polymerase exonuclease nickase dynamic network assembly (PENDNA)* toolbox [62], a molecular programming framework that can be used for several applications such as the rapid and very sensitive detection of biomarkers [63] or the detection of various activities for enzyme engineering [64]. The properties of the different DNA isothermal amplification techniques have been summarized in Table 1.

3.3 Applications of Cell-Free DNA Replication in Drops

3.3.1 Detection and Diagnostics Techniques

Microdroplets are compartments that can act as miniaturized independent reactors, with a volume ranging from hundreds of femtoliters to tens of nanoliters. This allows for highly parallelized biochemical reactions, manipulation of small objects (such as single cells), and detection of compounds at the single-molecule level. In such a small volume, a single molecule can reach a concentration of several femtomolars, which is suitable for most of the dynamic ranges of the abovementioned DNA amplification methods. Many probing strategies exist to visualize a product of amplification, with non-specific intercalating dyes being the most commonly used. The possibility to efficiently amplify a single copy in a droplet and detect the product of the reaction has paved the way for the implementation of absolute quantification techniques. By adjusting the target concentration to encapsulate, on average, around one molecule per droplet, it is possible to determine the proportion of droplets where the reaction occurred. Since the probability of the presence of a molecule in a droplet after encapsulation follows Poisson's law, one can easily calculate the concentration of the target in the original sample. This technique has been largely used to detect and quantify biomarkers in the fields of medical diagnostics and food security. Since its first implementation in tubes for the detection of a mutant *ras* oncogene related to colorectal cancer using PCR [65], many technological improvements have been made to adapt, develop, and generalize the use of this technique [66]. The first high-throughput droplet-based implementation for the absolute quantification of the *Vaccinia* virus genomic DNA [67] marked the beginning of a rapid spread of the use of this technique. Numerous reviews already list its many applications, including cancer diagnostics and prognostics [68–70], diagnostics of infectious diseases, [71–73], and food safety [74–76].

The first implementation of the droplet-based digital LAMP in an integrated chip demonstrated its high sensitivity for *Neisseria gonorrhoeae* DNA detection [77], and its utility for the entire domain of medical diagnostics. This method has proven to be efficient for the detection of other pathogenic bacteria, such as *Salmonella typhimurium* [78, 79] or *Mycobacterium tuberculosis* [80] but could also be used to detect more specifically antimicrobial resistant strains of *E. coli* [81] or vancomycin-resistant *enterococci* [82]. The method has also been adapted for the absolute quantification of viral DNA, such as hepatitis C virus [83], avian Flu H5N1 [84], and, in a multiplexed version, HIV [85]. Additionally, it has been used for cancer biomarkers, such as HPV16 [86] and DPAK1 [85], for cervical cancer, as well as c-Myc associated with 40% of all human tumors [87].

RCA and its derivatives (C2CA, HRCA, MDA) have been successfully used for droplet-based applications. For diagnostics, the digitalized version of RCA has been shown to be relevant for the detection of cancer cell surface protein marker [88]. Coupled with a DNAzyme, it was able to detect *E. coli* at a single-cell sensitivity in urine sample [89], with direct possibility of adaptation of the method to *Helicobacter pylori*, *Clostridioides difficile*, *Legionella pneumophila*, or *Staphylococcus aureus*. Using magnetic beads moving between droplets containing various reagents and targeting the DNA of *Pseudomonas aeruginosa*, C2CA was useful for ultra-sensitive infectious disease diagnostics [90]. HRCA, similarly to its counterpart MDA, suffers from more unspecific background amplification than the isothermal method previously cited; it could nevertheless be implemented for the digital quantification of circHIPK3, a circular RNA biomarker of several cancers [91]. This principle had already been demonstrated years before, without diagnosis aim, but for clonal DNA amplification [92]. In this study, they amplified compartmentalized single *lacZ*-encoding plasmids before merging each individual droplet containing an in vitro translation mix and a fluorescent substrate and before performing a digital counting. This work opened perspectives on the creation of fully in vitro systems for directed evolution. MDA has the peculiarity of being unspecific and presenting a problematic background amplification. The droplet-based digitalization offers the opportunity to drastically reduce the competition for resources between DNA fragments to be amplified and parasites, with applications for the whole-genome sequencing [93–95]. RPA, in its digitalized version, also demonstrated several successes in the medical diagnostics domain. It proved to efficiently target specific *loci* in the DNA of *S. aureus* [96], *C. difficile* [97], and *Listeria monocytogenes* [98] or in antibiotic resistance genes of *E. coli* [99].

PENDNA-based reactions have also been encapsulated into droplets. To explore several parameters of a biochemical reaction at once, millions of droplets containing different amounts of reagents were produced. The reactions in each compartment were followed under an epifluorescence microscope. It was possible to find the composition of the droplets presenting an interesting behavior thanks to fluorescent reporters of the concentrations of the reagents studied [100]. This use of droplets as miniaturized reactors to highly parallelize the study of reactions has recently been used for the implementation of enzymatic neural networks [101]. On the other hand, the progressive development of detection modules to convert small molecules concentration, biomarkers concentration, or enzymatic activity into processable oligonucleotides [63, 64] makes this molecular programming toolbox appealing for the development of new diagnostic techniques. A digital detection platform for microRNAs (miRNA), biomarkers of many cancers among others, has been implemented and proved to be highly sensitive [63]. The multiplexing of this technique to detect up to six miRNAs at once is currently being developed [102] in parallel with an extension to other biomarkers [103].

3.3.2 Directed Evolution of Polymerases

Droplet digital DNA amplification paved in fact the way for the creation of several ultra-high-throughput directed evolution platforms based on the principle of self-selection. In 2001, Philipp Holliger's laboratory reported their results regarding the directed evolution of the *Taq* polymerase [104]. In this technique, a random library of around 10^6 mutants of the DNA polymerase is cloned into bacteria that express the proteins. The bacteria are then encapsulated into w/o droplets with primers specific of the polymerase gene, PCR buffer, and dNTPs. When the emulsion is submitted to thermocycling, the active mutant polymerases replicate their own gene, whereas the inactive ones do not. After emulsion break, a new library, enriched with the best mutants, is retrieved. A comparison of the sequencing data extracted from the libraries before and after the selection allows the identification of the beneficial mutations.

The principle was implemented in the team of Andrew Ellington by replacing PCR by RCA to adapt the method to less thermostable strand-displacing enzymes [105]. The same team created an extension of this method in order to generalize it to non-replicative enzymes. They used *in cellulo* genetic circuits linking the activity of the targeted enzyme to the production of *Taq* polymerase and demonstrated their ability to make the T7 RNA polymerase and a tRNA-aminoacyl-synthetase evolve [106]. Moreover, to extend this principle with a controlled in vitro activity test of the targeted enzyme, the members of Yannick Rondelez's laboratory interfaced their PENDNA-based molecular programs with the replicative PCR [107]. By converting the enzyme activity into a quantification of primers produced, they could demonstrate a strong selectivity to improve the Nt.BstNBI endonuclease activity and thermostability.

3.3.3 Single-Cell and Omics Applications

The ability to compartmentalize DNA amplification in miniaturized reactors also represents an opportunity to massively parallelize single-cell processes. In 2017, Adam Abate's team developed a barcoding system that allowed for sequence single-cell genomes at an unprecedently high throughput [108]. Following a multi-re-encapsulation and droplet merging process, they were able to lyse the cells, fragment the genomic DNA, and PCR amplify the fragments with a unique barcode. After emulsion breakage, they sequenced all the retrieved DNA before reassembling the genome in silico. They applied this method to a pool of Gram-negative and Gram-positive bacteria and fungi for the analysis of the distribution of the antibiotic resistance genes, the virulence factors, and phage sequences across a microbial community. A year later, a very similar process was implemented in Dennis Eastburn's team to study the heterogeneity of acute myeloid leukemia tumor cells, opening new options for more personalized cancer treatments [109].

The same principle can also be applied to transcriptomics by parallelizing the cDNA production and barcoding from encapsulated individual cells' mRNA [110–112]. To quickly quantify the level of transcription of specific genes, a LAMP-based single-cell reverse transcription and amplification revealed to be particularly efficient to detect discrepancies in the expression, among different cell types, of the hydroxymethylbilane synthase gene, which is involved in neurodegenerative disease involved [113], or to detect variations in the expression of ERBB2 breast cancer marker gene [114].

4 Cell-Free Transcription of RNA

4.1 *RNA Transcription Cell-Free Systems*

Although their use is not as common as DNA amplification yet, cell-free transcription systems have started to gain importance in the field of droplet microfluidics. A recent application of the in vitro transcription lies in the context of the COVID-19 outbreak, with the necessity of an industrial production of mRNA for RNA-based vaccine technology newly applied to human health. Starting from the easily synthesized DNA gene, a DNA-dependent RNA polymerase, such as the T7 RNA polymerase, known for its high activity, has been used to produce large quantities of transcripts, subsequently purified and chemically modified [115]. The promising new version of these vaccines based on self-amplifying RNA (saRNA) is also based on the same principle regarding its industrial production [116]. In vitro transcription is similarly used for aptamer discovery and production [117]. The SELEX method for RNA aptamers selection requires indeed a reverse-transcription step, an error-prone DNA amplification, and a new in vitro transcription step between each selection by binding. Once selected, the binders are produced in large quantities by in vitro transcription, purified and chemically modified to increase, for example, their resistance to nucleases.

To amplify exponentially a specific RNA sequence, several extremely similar methods were created in the beginning of the 1990s, including self-sustained sequence replication (3SR) [118], nucleic acid sequence-based amplification (NASBA) [119], or transcription-mediated amplification (TMA) [119]. All these methods start from a single-stranded RNA target which is reverse-transcribed. The RNA is digested by the RNAse H and the resulting single-stranded DNA is then complemented using a primer presenting a tail that contains the T7 promoter. The T7 RNA polymerase can then produce several copies of the initial RNA target that can, in turn, be reverse transcribed. This exponential amplification has, for example, been used to implement a commercialized diagnostic kit for rapid detection of *M. tuberculosis* [120]. More recently, Ju et al. created the nicking and extension chain reaction system-based amplification (NESBA), coupling NASBA with SDA exponential DNA amplification [121]. This method allowed for an increase in sensitivity of two orders of magnitude for detecting the respiratory syncytial virus A (RSV A) and the SARS-CoV-2 genomic RNA [121, 122].

4.2 *Cell-Free RNA Transcription Applications in Drops*

4.2.1 Directed Evolution of RNA Molecules

During the last few years, research on the optimization of biologically relevant RNA molecules in drops has been focused, mainly, on ribozymes and RNA fluorogenic aptamers. On one side, the ability of the ribozymes to catalyze biological activities [123] makes them promising candidates to control several synthetic biology processes [124] after tuning their properties. On the other side, fluorogenic RNA aptamers can be used in various cell-related microscopies like confocal [125] or super-resolution [126] due to their notable brightness when labeled to small-molecule fluorogens, their ability to cross cell walls, and their small size compared to fluorescent proteins.

Due to their relatively new discovery compared to classical proteogenic enzymes and the difficulty to handle RNA samples, there are not many trials trying to improve the properties of ribozymes in vitro. However, findings related to these types of molecules increase every year. Thus, one can believe that the number of studies trying to optimize these molecules will rise in the next decade. One interesting study using a cell-free in vitro transcription mixture in drops optimizes the catalytic properties of an X-motif capable of RNA cleavage via an internal phosphoester transfer reaction [127]. For that purpose, PCR reagents and DNA mutants were encapsulated in drops first before an off-chip PCR. Next, two picoinjection steps were performed. In the first one, the T7 RNA polymerase-based transcription mix was injected. The picoinjection added a fluorogenic nuclease assay. The final sorting step based on the fluorescence signal, and therefore the activity of the ribozyme, of the drops gave rise to a new ribozyme with 28-fold enhanced catalytic properties after nine rounds of selection and several mutations.

Contrary to ribozymes, the cell-free directed evolution of aptamers has resulted in a larger number of optimized new molecules. Aptamers are biomolecules that can bind specifically to other molecules. Riboswitches (RNA aptamers) take part in several biological processes, including metabolite-dependent gene expression control [123]. Droplet-based microfluidic technologies enable the interrogation of different variants of these aptamers in each drop. In the case of fluorogenic aptamers, the sorting step can be easily performed with a FADS system. Fluorogenic aptamers are becoming now common in many biochemistry and synthetic biology fields and their optimization is a hot area of research. In this sense, cell-free RNA production systems have been used to improve several G-quadruplex RNA aptamers and their subsequent biosensors. In particular, the aptamer-based fluorogenic biosensor iSpinach [128, 129], the MangoIII aptamer [130, 131], and the Gemini-oCoral fluorogenic dimer [132] were developed. The microfluidic workflow used for the optimization of these biomolecules was similar for all the cases reviewed (Fig. 5): (1) encapsulation of a unique variant per drop, (2) amplification by droplet PCR (dPCR), (3) reinjection and fusion with in vitro transcription reagents, (4) incubation of the drops, and (5) sorting of the drops by FADS and posterior sequencing. It must be noted that the step iv can be performed either off- or on-chip.

4.2.2 Other Applications

Other fields of interest in transcribing RNA molecules in drops are the interactions of this RNA with organelles. For instance, Aufinger et al. [133] developed a method for

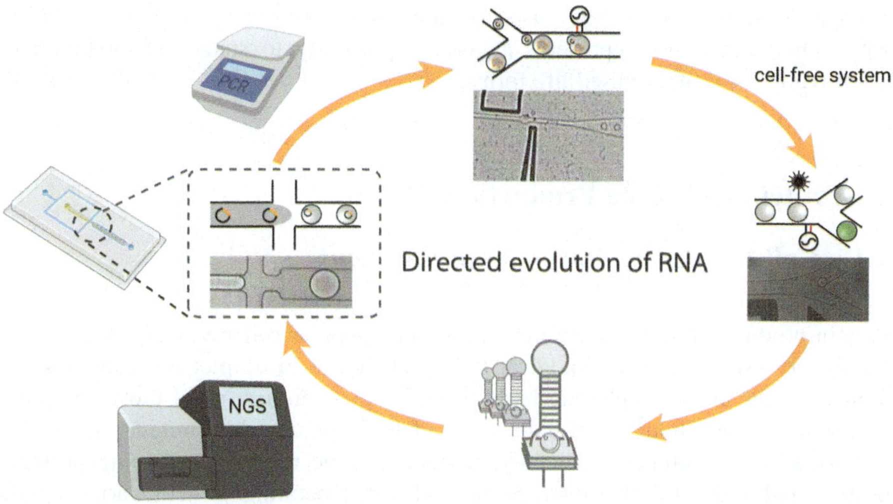

Fig. 5 Microfluidic workflow for the directed evolution of RNA aptamers and ribozymes. The process consists of three well-defined microfluidic steps: encapsulation, droplet fusion, and sorting. dPCR amplification, cell-free transcription incubation, and NGS sequencing are performed off-chip (Created with BioRender.com)

the covalent immobilization of DNA molecules onto agarose microbeads and their posterior transcription and translation inside water-in-oil drops. They demonstrated that mRNAs transcribed from transcription organelles can be exclusively directed to catch organelles via hybridization to the matching DNA addresses. They also proved that when controlled by toehold switch riboregulators, these mRNA molecules are exclusively translated in translation organelles comprising their equivalent DNA triggers. The development of transcriptional oscillators inside droplets has also attracted the interest of the scientific community in this area of research. Friedrich C. Simmel and collaborators developed a programmable transcriptional oscillator system with variations in the amplitude, frequency, and damping of the oscillator in microemulsions [134, 135]. The application of this technique to droplet microfluidic methods would enable the rapid and precise control of such oscillators.

Following a similar principle based on a reverse transcriptase, the RNAse H, a DNA polymerase, and an RNA polymerase, a molecular programming platform for RNA signal processing called RTRACS has been created [136]. This system can perform basic logic operations and can be used to model complex behaviors and program reactions to certain stimuli. It could be useful in many applications such as drug delivery systems. Additionally, the NESBA has recently been implemented in a droplet-based digital version for absolute quantification of the RSV A [122].

Finally, in the extension of their study on DNA self-replicators, the team of N. Ichihashi developed an RNA-based self-replicating system using the RNA-dependent RNA polymerase or RNA replicase from the Qβ bacteriophage. By encapsulating it into droplets, they could study the evolution of their RNA replicators and, among other aspects, how they coexist with parasites [137–139]. This framework represents a good opportunity to study the evolution of primitive RNA-protein-based life forms.

5 Protein Cell-Free Production Systems

5.1 Cell Lysates vs. Recombinant Systems

Protein production systems are among the most popular cell-free systems, and they are widely used in many biological fields. Their use in droplet microfluidics has numerous advantages including high-throughput screenings, rapid protein production within a few hours, and expression of toxic or unstable proteins, inter alia. Moreover, the combination of these systems can be used for screening protein–protein and protein–DNA interactions and for generating protein variants with improved or new functions by directed evolution. The number and diversity of cell-free protein production systems have grown remarkably during the last few years.

On one side, cell lysates extract all the molecular machinery necessary for protein production and combine them later with different energy regeneration systems

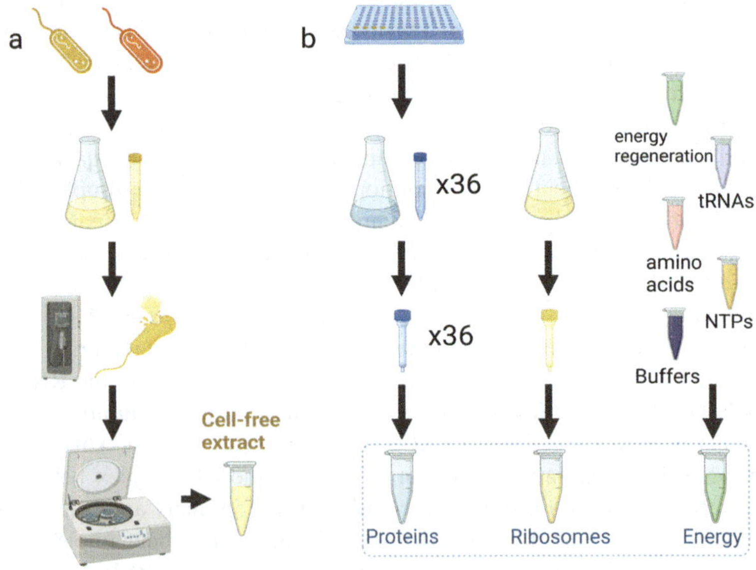

Fig. 6 Comparison between cell-free extracts and recombinant systems production. Panel (**a**) shows the relatively easy process of growing-lysis-clearing for producing cell-free extracts. Panel (**b**) displays the 36 necessary protein purifications to produce a recombinant cell-free system. Purified ribosome and an energy system must be also added thereafter (Created with BioRender.com)

(Fig. 6a). Historically, lysates belonging to various strains of *E. coli* have been the most commonly used. The choice of strain strongly depends on the application, with the BL-21 variants being the most frequently used [140, 141]. Nowadays, the lysate toolbox has been expanded to many different organisms, including yeasts, rabbit reticulocyte, and insects, among others. Advantages of using lysates in drops include higher yields and easiness to summarize entire cellular processes. The main disadvantages are high viscosities, variability between preparations, and imprecise compositions [141]. However, this issue can be addressed by using proteomic tools to control the batch [142] and strain variability [143].

On the other side, completely recombinant cell-free systems have the advantage that their composition can be tuned according to specific requirements (Fig. 6b). The use of these systems is therefore advantageous for the study of minimal systems and for building bottom-up artificial cells. Although attempts had been made in the past [144] the first high-yield recombinant system, the aforementioned PURE [7], was developed in 2001. The main disadvantage of these systems is their high cost, as well as variability between batches of the commercially available systems, translation rates, and yields [145]. In order to overcome these issues, the laboratory of Sebastian Maerkl has developed the OnePot PURE system [146, 147], where all the proteins (except the ribosomes) are prepared in only one co-culture and purification step.

5.2 *Applications of Cell-Free Protein Production Systems in Drops*

5.2.1 Protein-Directed Evolution

Although directed evolution of proteins is now a mature area of research and allows the user to link the genotype with its phenotypic properties, examples combining cell-free expression systems and droplet microfluidics are still scarce. The seminal work developed by Dan Tawfik and Andrew Griffiths in 1998 with the directed evolution in drops HaeIII methyltransferase [148] followed by the work of Tawfik on protein inhibitors of DNA nucleases [149], the one of Griffiths on beta-galactosidases [150], and an application to biotin protein ligase by Ellington and coworkers [151] suggested a huge development for the field in the following years. However, since then, the field has evolved to cytoplasmic [22, 23, 27, 29, 152] and cell surface display [153] screens and selection methods, and only a few examples of fully in vitro systems are available in bibliography.

Among the advances toward these fully in vitro methodologies, selecting enzymes or ligands in well-controlled conditions, the development of microfluidic chips interfacing fluidic channels with electrodes represented a good opportunity to extend Griffiths' in vitro compartmentalization (IVC) to a wider range of targets. The creation of FADS [19], which enables to sort droplets according to their levels of fluorescence thanks to dielectrophoretic forces, and the conception of the picoinjection or of the droplets merging modules, which allow adding reagents to a droplet after its initial formation, revealed to be highly versatile for fluorescence-producing reactions [92]. This kind of platform proved its feasibility with the successful evolution of a beta-galactosidase [154].

One of the most active groups in this area is the laboratory of Florian Hollfelder. They have recently reported the directed evolution of the protease *Savinase*, which is toxic for *E. coli* and, therefore, cannot be properly expressed [8]. The system consists of a three-step microfluidic workflow: DNA rolling-circle amplification, IVTT reagents picoinjection, and a second picoinjection to add the fluorescent substrate. Droplets were later sorted based on the protease activity. They optimized the initial protease activity up to fivefold with only two screening rounds. Another interesting enzyme optimization using CFPS was performed by Sakatani and coworkers with the phi29 DNA polymerase [155]. Using a similar microfluidics workflow (compartmentalization, rolling-cycle replication, and gene expression) they performed a completely in vitro evolution of the polymerase under six different conditions. Under all the conditions, they found various mutations that enhanced the rolling-cycle amplification by the encoded polymerase, especially the double mutant K555T/D570N.

For the proteins that can be expressed from a single DNA molecule, several techniques have been created to directly link the proteins of interest to their gene. With this so-called gene display method, single genes are encapsulated in droplets containing an in vitro expression mixture. The resulting proteins can then be attached to the DNA molecule, before the emulsion breakage and the recuperation of the good mutants. This methodology has been successfully implemented for the evolution of the HaeIII methyltransferase that could bind covalently its own DNA molecule through a 5-fluoro-2′-deoxycytidine base [156]. The team led by Hollfelder implemented a SNAP-tag-based strategy where the gene carrying benzylguanine moieties at its ends is covalently bound by the SNAP polypeptide fused with the protein of interest. This technique was used for the development of DARPin binders targeting the Her2 transmembrane protein, and allowed for a 10^7-fold enrichment of a good binder over an unselected DARPin in three rounds [157].

Another interesting methodology to maintain genotype–phenotype linkage in a fully in vitro context is the use of beads to display the activity of certain reactions in drops. Often, a proper expression yield cannot be obtained from a single copy of the gene. A microbead can play the role of a physical support for monoclonal amplification. The beads can then be encapsulated in a cell-free expression mixture where the proteins of interest are produced before getting attached to the bead through a physical link such as an affinity tag or a covalent bond. Such methods are of particular interest when the amplification, expression, and selection steps are incompatible and require washing steps. The main drawback being the necessary multiple encapsulations that reduce the throughput of the method. In this context, Diamante et al. implemented a workflow based on streptavidin-coated magnetic beads encapsulated with a single copy of the gene of interest. Thanks to 5′-biotinylated primers, each mutant gene could be amplified by PCR and attached to its bead. After emulsion break, wash, and reencapsulation in an in vitro transcription and translation, the multiple copies of each gene could trigger the production of HA-tag variants fused with a SNAP-tag that created a covalent bond with a benzylguanine substrate coupled with a biotin. After a new emulsion breakage, the beads displaying each both several copies of a variant of the gene of interest and several of the corresponding proteins could be submitted to the discrimination process that consisted in this case in the binding of a fluorescent dye-labeled anti-HA antibody and a sorting by FACS [158].

Later, the team of Yannick Rondelez created their own bead display platform for the evolution of the toxic Nb.BsmI nicking endonuclease. The beads were produced using microfluidics and were based on a biocompatible hydrogel. They contained a circularized gene and the material to perform a monoclonal HRCA. The beads were then re-encapsulated in a cell-extract-based protein expression mixture where the nicking enzymes fused to a SNAP-tag were produced and attached inside the gel matrix displaying the benzylguanine substrate. After emulsion breakage and a thorough wash, the beads could be finally encapsulated with a PENDNA-based activity assessment module linked to the production of the primers necessary for subsequent PCR [159]. In parallel, the team of Hollfelder implemented a similar platform based on a different hydrogel, which they used for the evolution of DARPin binders and LL37 antimicrobial peptides [160]. In a recent publication, they demonstrated the adaptability of their platform to IgG, SUMO-tag, Spy-tag, and Snp-tag binding technologies [161].

5.2.2 Expression Improvement and Biomolecular Circuits

Attempts to improve the protein expression levels of CFPS systems in droplets have also been made. For instance, Bashaw et al. analyzed the expression efficiency levels of 19 different bicistronic designs (BCDs). BCDs are DNA constructs with two ribosome binding sites (RBS), where the first one translates a leader polypeptide with no biological relevance that can direct the production levels of the second protein [162]. They reported that three of these designs have higher and more predictable translation efficiencies, making them more suitable for protein engineering applications than the classical on RBS systems. Akin to this piece of research, Sierra et al. studied how to enhance multi-gene expression levels with cell-free systems [163]. They encapsulated multicistronic DNA molecules together with CFPS systems to track the expression efficiency of the corresponding genes. The results highlighted that it is possible to express multiple genes in drops with in vitro systems, both with the DNA in solution or when it is conjugated to a microbead.

One more application of CFPS of proteins in droplets is the development of biomolecular circuits [164]. Using cell-free systems and droplet microfluidics, the lab of Adam Abate constructed a platform that modulates the biocircuit parameters, avoiding the exhausting cloning process used for circuit optimization. Using microfluidic parameter space scanning allows testing large combinatorial libraries and can also be used for other biologically relevant applications.

5.2.3 Protein Interactions

The molecular communications of proteins with other proteins (protein–protein interactions), DNA (protein–DNA interactions), and even bigger molecular machines such as ribosomes or liposomes have also been studied by the combination of CFPS and droplet microfluidics. The use of these systems has several advantages compared to cell-based assays, such as the lack of undesired or non-specific interactions with other cellular materials or the possibility to tune the composition of the CFPSs. For instance, the combination of CFPSs and droplet microfluidic techniques is currently being optimized to screen the activity of antimicrobial peptides (AMPs) and develop new ones to fight the surge of antimicrobial resistance, one of the main global health issues in the next decades. Charon et al. proposed a method to discover new ribosome inhibitory peptides by encapsulating a bicistronic DNA molecule encoding a peptide library and the GFP and a posterior picoinjection of an in vitro transcription and translation system [165]. If the initial peptide inhibits the ribosome, the GFP would not be translated, and vice versa, if the peptide is inhibitory, the GFP would be expressed (Fig. 7a). The less fluorescent drops would be finally sorted. This approach, therefore, takes the ribosome as a discovery platform for new AMPs by quantifying peptide–ribosome interactions. Another mechanism used by AMPs is based on their capacity to disrupt selectively the bacterial cell membrane because of their cationic nature [166]. Most AMPs exhibit a net positive charge due to the large quantity of Lys and Arg amino acids in their structure. Thus, they will preferentially interact with negatively charged membranes. Based on this premise, the lab of Petra S. Dittrich has created a droplet microfluidic platform to screen the interactions of AMPs with both positively and negatively charged liposomes [167]. They co-encapsulated into the same double emulsion drop a cell-free extract, DNA templates, and liposomes containing different fluorescent dyes depending on their positive or negative nature (Fig. 7b). The extract expresses the AMP encoded in the DNA and the peptide can immediately interact with the liposome. The activity of the AMP is measured by quantifying the leakage of each dye visually with a microscope and by means of a flow cytometer.

5.2.4 Antibody Engineering

The combination of cell-free systems and droplet microfluidics has also been used in antibody development and research since it is possible to produce multiple antibodies and, therefore, offers a large library capacity. Sun et al. wrote a general review of how microfluidic technologies can accelerate the antibody discovery process [168]. Nevertheless, this technology is not yet mature and only a few examples are available in recent bibliography. Jacková et al. developed a DNA-encoded immunoassay in picoliter drops to study the molecular recognition between antigens and antibodies [169]. They co-encapsulated a DNA coding for the variable domain of the heavy chain of heavy-chain antibodies (VHH), PURE

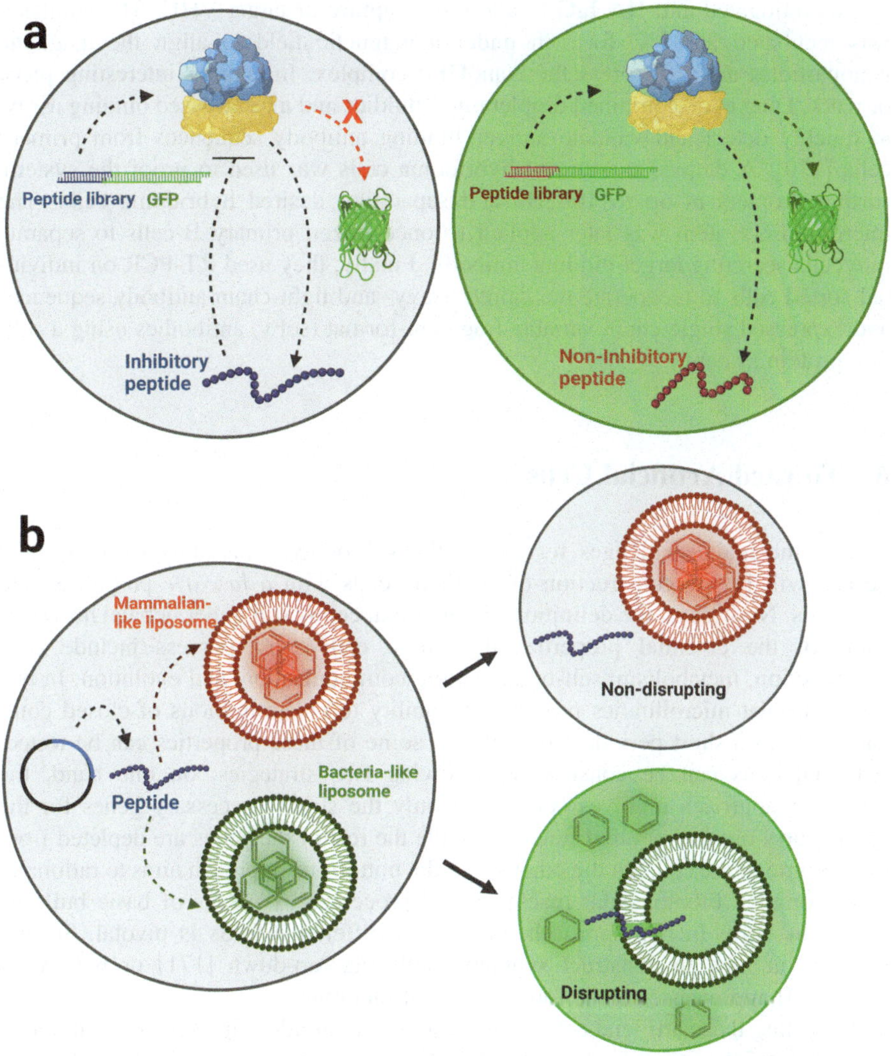

Fig. 7 Different strategies to screen antimicrobial peptides in drops with CFPS systems. Panel (**a**) shows the approach of using the ribosome as a discovery platform for ribosome inhibitory peptides that uses a GFP protein as an inverse reporter of the inhibiting activity. Panel (**b**) displays the strategy of using CFPS systems to produce cell-wall disrupting peptides and the multiplexing with two pairs of liposome-dye pairs (Created with BioRender.com)

system, the target antigen, and a capture scaffold where the antigen–antibody complex accumulates to screen the fluorescence signal of the molecular recognition. The fluorescent reporter is an in vitro synthesized, functional hemagglutinin (HA)-tagged anti-GFP VHH (named NanoGFP). The capture scaffold is composed of several streptavidin-coated magnetic nanoparticles that were functionalized with biotin-conjugated anti-HA IgG to allow the capture of target VHH. The emulsion was incubated at 37°C for 3 h under a magnetic field to align the magnetic nanoparticles and to express the NanoGFP complex. In another interesting piece of work, Ding et al. combined droplet microfluidics and a bead-based binding assay, to quickly detect and validate antigen-binding antibody sequences from primary cells [170]. A defined mixture of hybridoma cells was used to proof the system, sorting droplets at up to 100 Hz and separating desired hybridoma cells. The microfluidic system was later applied to once-frozen primary B-cells to separate rare cells secreting target-binding antibody. Finally, they used RT-PCR on individual sorted cells to recuperate the paired heavy- and light-chain antibody sequences and expressed single-chain variable fragment-format (scFv) antibodies using a cell-free protein system.

6 Toward Artificial Cells

One of the main challenges for the synthetic biology community over the next decade will be the construction of synthetic cells with *a la carte* properties and functions. Nowadays, the definition of a minimal cell is still under debate; however, some of the essential properties that these cells must possess include self-reproduction, metabolism, self-organization, communication, and evolution. In this sense, droplet microfluidics provides the ability to create millions of closed compartments in a short period of time where some of these properties can be tested. Artificial cells can be constructed following two strategies: on one hand, the top-down approach involves conserving only the strictly necessary genes for the cell to carry out the required function, while the rest of the genes are depleted from the genome of the cell. On the other hand, the bottom-up approach aims to rationally build complex biomolecular machines and processes by means of basic building blocks in vitro. In this approach, the use of different CFPSs is pivotal. Several reviews on how to construct synthetic cells via top-down [171] or bottom-up [172, 173] approaches can be found in recent literature.

Providing these artificial cells with the ability to divide (Fig. 8a) has been one of the targets of the scientific community in this field. Godino et al. explored the min biochemical network in vitro into liposomes, responsible for the regulation of bacterial cell division [174]. This system is an example of a self-organizing molecular system. They encapsulated the DNAs encoding the MinE and MinD proteins together with the PURE system into liposomes to study their dynamic protein patterns. They found that these proteins form stable patterns that also drive liposome deformations. When this system is integrated with the cytoskeletal proteins FtsA and FtsZ, min is also able to dynamically regulate the FtsZ patterns. In this context, the bacterial susceptibility of antibiotics in drops regarding their ability to divide has been also studied [175]. Using a droplet array system, they screened the proliferation of *E. coli* for different concentrations of antibiotics by fluorescent and light scattering measurements. This system also contributes to the microfluidic automation of

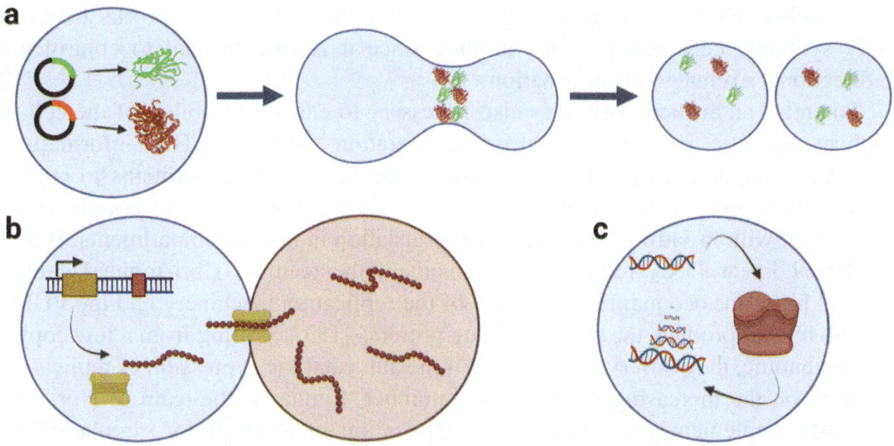

Fig. 8 Three of the properties that artificial cells must have in droplet systems discussed in this review. (**a**) Cell division. Artificial must be able to self-divide with the help of proteins responsible for this task in cellular processes. (**b**) Cell communication: artificial cells must be able to send and receive information from their counterparts. (**c**) Self-replication. Artificial cells must be able to self-replicate fundamental genetic information before proliferation to assure the continuity of the lineage (Created with BioRender.com)

large-scale antibiotic interactions screens and single-cell resolved populations resistance studies.

One of the properties that artificial cells should have is the ability to communicate between them (Fig. 8b). The combination of droplet microfluidics and CFPS systems allows the study of this communication in a basic way, since it is possible to generate protocells capable of processing fundamental gene instructions. In this manner, it is possible to achieve specific and programable communication between two drops. For instance, CFPSs have been used to program specific communication between two drops with a signaling molecule using LacY transporter proteins [176]. Similar approaches have been developed using liposomes as emitters and proteinosomes as receivers [177] and between synthetic mammalian and bacterial protocells [178] (using HeLa and *E. coli* extracts, respectively). Communication can also be achieved through amphiphilic molecules that can be present in both the water and oil phase and thus can be transported through a micellization process [179]. Droplets with a lipid monolayer that acts as a glue have been also tested [180]. This system can adhere several droplets together in two or three dimensions by creating a lipid bilayer that could act as the messenger platform between drops. This research also has implications for the previously mentioned self-organization requisite of artificial cells. In another interesting piece of work, Dupin and Simmel used this system with various types of feedforward, and feedback in vitro gene circuits, to implement artificial cell signaling processes [181]. The same author and collaborators were also able to extract the positional information of the drops through their morphogen gradients [182]. All this research shows the potential of droplets to act as suitable compartments for protocell communication despite some disadvantages such as poor permeability. However, engineering collective communication networks between fully synthetic cells is still at its infancy since it is still difficult to control the bidirectional exchange of information.

To create an artificial cell, it is also necessary to ensure the ability of the cell to replicate genetic information before proliferation (Fig. 8c). This information, encoded in nucleic acid sequences, needs to be translated into proteins to realize cellular functions. In this context, several teams have attempted to couple DNA replication with in vitro transcription and translation in micro-compartments. Based on Phi29 linear DNA replication mechanism, the team of Christophe Danelon created liposomes containing the genes of the replication machinery and the PURE system able to produce the corresponding proteins [183]. Starting from a few copies of the genome, they observed the effect of the autocatalytic expression of replicative proteins on the increasing genome copy number. Similarly, the team of Norikazu Ichihashi implemented an RCA-based replication mechanism for circular DNA encoding the replicative polymerase and a recombinase (to circularize the copied DNA) in w/o droplets [184, 185]. To mimic cell divisions, they mixed the droplets with empty fresh droplets every 16 h and assured the exchange of fresh reagents and genetic material in the new population through mechanical mixing. This way, they could follow the evolution of 30 generations and notice that the DNA accumulated common mutations and exhibited a higher replication ability. They could also demonstrate the possibility to create a similar self-reproductive system in which the tRNAs, essential for the translation step, are not present in the initial translation mix but encoded in the replicating DNA [186]. In addition to the fundamental interest that present these self-replicative systems for the synthetic biology community, the same team also proved the possibility to use these systems for the fully in vitro directed evolution of the Phi29 DNA polymerase [155]. However, it must be noted that these self-replicating systems remain quite limited regarding the size of the genome to be replicated. To overcome this limitation, the team of Hiroyuki Noji implemented in droplets a fully reconstituted replisome of *E. coli* to amplify large (more than 100 kb) genome from a single copy, paving the way for very complex artificial cells carrying a large number of genes [187].

7 Concluding Remarks

In this chapter, we aim to review the state of the art of the cell-free macromolecular synthesis toolbox and many of its applications in droplet microfluidics. DNA amplification methods have already reached maturity in terms of the types of processes and have been shown to be a real alternative with clear advantages compared to host organism-based replication processes, especially in the field of medical diagnosis and detection. In contrast, the transcriptome and translatome, which are less understood and characterized, are still in their initial-mid stages of technological transfer to industrial and health applications. The following decade will be crucial for these technologies, and some expected scientific milestones, such as the universalization of protein sequencing [188], will speed up the number of applications and industrialization processes.

Moreover, to further increase the throughput of the directed evolution, while keeping the advantages of the cell-free processes, the creation of methods allowing the encapsulation of single genes in a mix where both the amplification and the expression could occur would represent a good opportunity to make processes easier, more robust, and faster. Such technology could have repercussions in other domains of protein sciences, such as functional metagenomics and its many prospective applications [189, 190].

Finally, efforts are being made for the bottom-up engineering of fully artificial cells. The use of droplet microfluidics as semipermeable compartments to test novel properties of these cells will help to accelerate the future applications of these biomolecular machines. Consortiums already working, such as MaxSynBio [191] from the Max Plack Society, Build-a-cell [192], or the European Synthetic Cell Initiative, will foster novel research lines in this field.

Acknowledgments A.L.C. is financed by Grant PID2019-111649RB-I00 and Grant PDC2021-120957-I00 funded by MCIN/AEI/10.13039/501100011033 and by the "European Union NextGenerationEU/PRTR." A.M. is financed by grant 2022-FELL-000011-01 funded by Gipuzkoa Fellows Program (Diputación Foral de Gipuzkoa) and grant "EPINPOC" co-funded by AECT Euroregion New Aquitaine-Navarra-Basque Country.

References

1. Hufnagel K, Reininger D, Ng SW, Gassert N, Rohland JK, Shahryarhesami S, Bauer AS, Waterboer T, Hoheisel JD (2019) In situ, cell-free protein expression on microarrays and their use for the detection of immune responses. Bio-Protoc 9:e3152–e3152
2. Kilb N, Herz T, Burger J, Woehrle J, Meyer PA, Roth G (2019) Protein microarray copying: easy on-demand protein microarray generation compatible with fluorescence and label-free real-time analysis. Chembiochem 20:1554–1562
3. Kopniczky MB, Canavan C, McClymont DW, Crone MA, Suckling L, Goetzmann B, Siciliano V, MacDonald JT, Jensen K, Freemont PS (2020) Cell-free protein synthesis as a prototyping platform for mammalian synthetic biology. ACS Synth Biol 9:144–156
4. Rhea KA, McDonald ND, Cole SD, Noireaux V, Lux MW, Buckley PE (2022) Variability in cell-free expression reactions can impact qualitative genetic circuit characterization. Synth Biol 7:ysac011
5. Mullis KB (1994) The polymerase chain reaction. Springer
6. Sridharan H, Piorino F, Styczynski MP (2022) Systems biology-based analysis of cell-free systems. Curr Opin Biotechnol 75:102703
7. Shimizu Y, Inoue A, Tomari Y, Suzuki T, Yokogawa T, Nishikawa K, Ueda T (2001) Cell-free translation reconstituted with purified components. Nat Biotechnol 19:751–755
8. Holstein JM, Gylstorff C, Hollfelder F (2021) Cell-free directed evolution of a protease in microdroplets at ultrahigh throughput. ACS Synth Biol 10:252–257
9. Swank Z, Laohakunakorn N, Maerkl SJ (2019) Cell-free gene-regulatory network engineering with synthetic transcription factors. Proc Natl Acad Sci 116:5892–5901

10. Manteca A, Gadea A, Van Assche D, Cossard P, Gillard-Bocquet M, Beneyton T, Innis CA, Baret J-C (2021) Directed evolution in drops: molecular aspects and applications. ACS Synth Biol 10:2772–2783
11. Baret J-C (2012) Surfactants in droplet-based microfluidics. Lab Chip 12:422–433
12. Chowdhury MS, Zheng W, Kumari S, Heyman J, Zhang X, Dey P, Weitz DA, Haag R (2019) Dendronized fluorosurfactant for highly stable water-in-fluorinated oil emulsions with minimal inter-droplet transfer of small molecules. Nat Commun 10:4546
13. Chowdhury MS, Zheng W, Singh AK, Ong ILH, Hou Y, Heyman JA, Faghani A, Amstad E, Weitz DA, Haag R (2021) Linear triglycerol-based fluorosurfactants show high potential for droplet-microfluidics-based biochemical assays. Soft Matter 17:7260–7267
14. Garstecki P, Fuerstman MJ, Stone HA, Whitesides GM (2006) Formation of droplets and bubbles in a microfluidic T-junction – scaling and mechanism of break-up. Lab Chip 6:437–446
15. Ward T, Faivre M, Abkarian M, Stone HA (2005) Microfluidic flow focusing: drop size and scaling in pressure versus flow-rate-driven pumping. Electrophoresis 26:3716–3724
16. Abate AR, Hung T, Mary P, Agresti JJ, Weitz DA (2010) High-throughput injection with microfluidics using picoinjectors. Proc Natl Acad Sci 107:19163–19166
17. Priest C, Herminghaus S, Seemann R (2006) Controlled electrocoalescence in microfluidics: targeting a single lamella. Appl Phys Lett 89:134101
18. Berry SB, Lee JJ, Berthier J, Berthier E, Theberge AB (2019) Droplet incubation and splitting in open microfluidic channels. Anal Methods 11:4528–4536
19. Baret J-C, Miller OJ, Taly V, Ryckelynck M, El-Harrak A, Frenz L, Rick C, Samuels ML, Hutchison JB, Agresti JJ (2009) Fluorescence-activated droplet sorting (FADS): efficient microfluidic cell sorting based on enzymatic activity. Lab Chip 9:1850–1858
20. Schmid L, Weitz DA, Franke T (2014) Sorting drops and cells with acoustics: acoustic microfluidic fluorescence-activated cell sorter. Lab Chip 14:3710–3718
21. Sakai T, Kamogawa K, Nishiyama K, Sakai H, Abe M (2002) Molecular diffusion of oil/water emulsions in surfactant-free conditions. Langmuir 18:1985–1990
22. Fischlechner M, Schaerli Y, Mohamed MF, Patil S, Abell C, Hollfelder F (2014) Evolution of enzyme catalysts caged in biomimetic gel-shell beads. Nat Chem 6:791
23. Larsen AC, Dunn MR, Hatch A, Sau SP, Youngbull C, Chaput JC (2016) A general strategy for expanding polymerase function by droplet microfluidics. Nat Commun 7:11235
24. Yan J, Bauer W-AC, Fischlechner M, Hollfelder F, Kaminski CF, Huck WT (2013) Monodisperse water-in-oil-in-water (W/O/W) double emulsion droplets as uniform compartments for high-throughput analysis via flow cytometry. Micromachines 4:402–413
25. Körfer G, Pitzler C, Vojcic L, Martinez R, Schwaneberg U (2016) In vitro flow cytometry-based screening platform for cellulase engineering. Sci Rep 6:1–12
26. Körfer G, Besirlioglu V, Davari MD, Martinez R, Vojcic L, Schwaneberg U (2022) Combinatorial in vitro flow-assisted mutagenesis (CombIMut) yields a 41-fold improved CelA2 cellulase. Biotechnol Bioeng 119:2076–2087
27. Agresti JJ, Antipov E, Abate AR, Ahn K, Rowat AC, Baret J-C, Marquez M, Klibanov AM, Griffiths AD, Weitz DA (2010) Ultrahigh-throughput screening in drop-based microfluidics for directed evolution. Proc Natl Acad Sci 107:4004–4009
28. Kintses B, Hein C, Mohamed MF, Fischlechner M, Courtois F, Lainé C, Hollfelder F (2012) Picoliter cell lysate assays in microfluidic droplet compartments for directed enzyme evolution. Chem Biol 19:1001–1009
29. Ma F, Chung MT, Yao Y, Nidetz R, Lee LM, Liu AP, Feng Y, Kurabayashi K, Yang G-Y (2018) Efficient molecular evolution to generate enantioselective enzymes using a dual-channel microfluidic droplet screening platform. Nat Commun 9:1–8
30. Karamitros CS, Lim J, Konrad M (2014) An Amplex red-based fluorometric and spectrophotometric assay for L-asparaginase using its natural substrate. Anal Biochem 445:20–23
31. Ostafe R, Prodanovic R, Lloyd Ung W, Weitz DA, Fischer R (2014) A high-throughput cellulase screening system based on droplet microfluidics. Biomicrofluidics 8:41102

32. Maceiczyk RM, Hess D, Chiu FWY, Stavrakis S, deMello AJ (2017) Differential detection photothermal spectroscopy: towards ultra-fast and sensitive label-free detection in picoliter & femtoliter droplets. Lab Chip 17:3654–3663
33. Choi JW, Kang DK, Park H, Demello AJ, Chang SI (2012) High-throughput analysis of protein-protein interactions in picoliter-volume droplets using fluorescence polarization. Anal Chem 84:3849–3854. https://doi.org/10.1021/ac300414g
34. Liu X, Painter RE, Enesa K, Holmes D, Whyte G, Garlisi CG, Monsma FJ, Rehak M, Craig FF, Smith CA (2016) High-throughput screening of antibiotic-resistant bacteria in picodroplets. Lab Chip 16:1636–1643
35. Goto H, Kanai Y, Yotsui A, Shimokihara S, Shitara S, Oyobiki R, Fujiwara K, Watanabe T, Einaga Y, Matsumoto Y (2020) Microfluidic screening system based on boron-doped diamond electrodes and dielectrophoretic sorting for directed evolution of NAD (P)-dependent oxidoreductases. Lab Chip 20:852–861
36. Smith CA, Li X, Mize TH, Sharpe TD, Graziani EI, Abell C, Huck WTS (2013) Sensitive, high throughput detection of proteins in individual, surfactant-stabilized picoliter droplets using nanoelectrospray ionization mass spectrometry. Anal Chem 85:3812–3816
37. Holland-Moritz DA, Wismer MK, Mann BF, Farasat I, Devine P, Guetschow ED, Mangion I, Welch CJ, Moore JC, Sun S (2020) Mass activated droplet sorting (MADS) enables high-throughput screening of enzymatic reactions at nanoliter scale. Angew Chem Int Ed 59:4470–4477
38. Wang X, Ren L, Su Y, Ji Y, Liu Y, Li C, Li X, Zhang Y, Wang W, Hu Q, Han D, Xu J, Ma B (2017) Raman-activated droplet sorting (RADS) for label-free high-throughput screening of microalgal single-cells. Anal Chem 89:12569–12577. https://doi.org/10.1021/acs.analchem.7b03884
39. Nitta N, Sugimura T, Isozaki A, Mikami H, Hiraki K, Sakuma S, Iino T, Arai F, Endo T, Fujiwaki Y (2018) Intelligent image-activated cell sorting. Cell 175:266–276
40. Isozaki A, Nakagawa Y, Loo MH, Shibata Y, Tanaka N, Setyaningrum DL, Park J-W, Shirasaki Y, Mikami H, Huang D (2020) Sequentially addressable dielectrophoretic array for high-throughput sorting of large-volume biological compartments. Sci Adv 6:eaba6712
41. Isozaki A, Harmon J, Zhou Y, Li S, Nakagawa Y, Hayashi M, Mikami H, Lei C, Goda K (2020) AI on a chip. Lab Chip 20:3074–3090
42. Mullis KB (1994) The polymerase chain reaction (Nobel lecture). Angew Chem Int Ed Engl 33:1209–1213
43. Dorak MT (2007) Real-time PCR. Taylor & Francis
44. Bachman J (2013) Reverse-transcription PCR (rt-PCR). In: Methods in enzymology. Elsevier, pp 67–74
45. Cirino PC, Mayer KM, Umeno D (2003) Generating mutant libraries using error-prone PCR. Dir Evol Libr Creat Methods Protoc:3–9
46. Notomi T, Okayama H, Masubuchi H, Yonekawa T, Watanabe K, Amino N, Hase T (2000) Loop-mediated isothermal amplification of DNA. Nucleic Acids Res 28:e63–e63
47. Lizardi PM, Huang X, Zhu Z, Bray-Ward P, Thomas DC, Ward DC (1998) Mutation detection and single-molecule counting using isothermal rolling-circle amplification. Nat Genet 19:225–232
48. Cao A, Zhang C (2012) Sensitive and label-free DNA methylation detection by ligation-mediated hyperbranched rolling circle amplification. Anal Chem 84:6199–6205
49. Dahl F, Banér J, Gullberg M, Mendel-Hartvig M, Landegren U, Nilsson M (2004) Circle-to-circle amplification for precise and sensitive DNA analysis. Proc Natl Acad Sci 101:4548–4553
50. Dean FB, Hosono S, Fang L, Wu X, Faruqi AF, Bray-Ward P, Sun Z, Zong Q, Du Y, Du J (2002) Comprehensive human genome amplification using multiple displacement amplification. Proc Natl Acad Sci 99:5261–5266
51. Silander K, Saarela J (2008) Whole genome amplification with Phi29 DNA polymerase to enable genetic or genomic analysis of samples of low DNA yield. Genomics Protoc:1–18

52. Vincent M, Xu Y, Kong H (2004) Helicase-dependent isothermal DNA amplification. EMBO Rep 5:795–800
53. Piepenburg O, Williams CH, Stemple DL, Armes NA (2006) DNA detection using recombination proteins. PLoS Biol 4:e204
54. Blanco L, Salas M (1985) Replication of phage phi 29 DNA with purified terminal protein and DNA polymerase: synthesis of full-length phi 29 DNA. Proc Natl Acad Sci 82:6404–6408
55. Hamdan SM, Richardson CC (2009) Motors, switches, and contacts in the replisome. Annu Rev Biochem 78:205–243
56. Sanders GM, Dallmann HG, McHenry CS (2010) Reconstitution of the B. subtilis replisome with 13 proteins including two distinct replicases. Mol Cell 37:273–281
57. Su'etsugu M, Takada H, Katayama T, Tsujimoto H (2017) Exponential propagation of large circular DNA by reconstitution of a chromosome-replication cycle. Nucleic Acids Res 45: 11525–11534
58. Lewis JS, Spenkelink LM, Jergic S, Wood EA, Monachino E, Horan NP, Duderstadt KE, Cox MM, Robinson A, Dixon NE (2017) Single-molecule visualization of fast polymerase turnover in the bacterial replisome. elife 6:e23932
59. Yeeles JT, Deegan TD, Janska A, Early A, Diffley JF (2015) Regulated eukaryotic DNA replication origin firing with purified proteins. Nature 519:431–435
60. Walker GT, Fraiser MS, Schram JL, Little MC, Nadeau JG, Malinowski DP (1992) Strand displacement amplification – an isothermal, in vitro DNA amplification technique. Nucleic Acids Res 20:1691–1696
61. Tan E, Erwin B, Dames S, Ferguson T, Buechel M, Irvine B, Voelkerding K, Niemz A (2008) Specific versus nonspecific isothermal DNA amplification through thermophilic polymerase and nicking enzyme activities. Biochemistry 47:9987–9999
62. Baccouche A, Montagne K, Padirac A, Fujii T, Rondelez Y (2014) Dynamic DNA-toolbox reaction circuits: a walkthrough. Methods 67:234–249
63. Gines G, Menezes R, Nara K, Kirstetter A-S, Taly V, Rondelez Y (2020) Isothermal digital detection of microRNAs using background-free molecular circuit. Sci Adv 6:eaay5952. https://doi.org/10.1126/sciadv.aay5952
64. Shenshin VA, Lescanne C, Gines G, Rondelez Y (2021) A small-molecule chemical interface for molecular programs. Nucleic Acids Res 49:7765–7774. https://doi.org/10.1093/nar/gkab470
65. Vogelstein B, Kinzler KW (1999) Digital pcr. Proc Natl Acad Sci 96:9236–9241
66. Sreejith KR, Ooi CH, Jin J, Dao DV, Nguyen N-T (2018) Digital polymerase chain reaction technology–recent advances and future perspectives. Lab Chip 18:3717–3732
67. Beer M, Reimann I, Hoffmann B, Depner K (2007) Novel marker vaccines against classical swine fever. Vaccine 25:5665–5670
68. Perkins G, Lu H, Garlan F, Taly V (2017) Droplet-based digital PCR: application in cancer research. Adv Clin Chem 79:43–91
69. Liao Y, Chen Y, Kou X, Xiao Y, Ye J, Wu A (2020) Diagnostic test accuracy of droplet digital PCR for the detection of EGFR mutation (T790M) in plasma: systematic review and meta-analysis. Clin Chim Acta 503:190–196
70. Olmedillas-López S, Olivera-Salazar R, García-Arranz M, García-Olmo D (2022) Current and emerging applications of droplet digital PCR in oncology: an updated review. Mol Diagn Ther 26:61–87
71. Li H, Bai R, Zhao Z, Tao L, Ma M, Ji Z, Jian M, Ding Z, Dai X, Bao F (2018) Application of droplet digital PCR to detect the pathogens of infectious diseases. Biosci Rep 38
72. Chen B, Jiang Y, Cao X, Liu C, Zhang N, Shi D (2021) Droplet digital PCR as an emerging tool in detecting pathogens nucleic acids in infectious diseases. Clin Chim Acta 517:156–161
73. Ishak A, AlRawashdeh MM, Esagian SM, Nikas IP (2021) Diagnostic, prognostic, and therapeutic value of droplet digital PCR (ddPCR) in COVID-19 patients: a systematic review. J Clin Med 10:5712

74. Qian C, Wang R, Wu H, Ping J, Wu J (2018) Recent advances in emerging DNA-based methods for genetically modified organisms (GMOs) rapid detection. TrAC Trends Anal Chem 109:19–31
75. Gao R, Liao X, Zhao X, Liu D, Ding T (2021) The diagnostic tools for viable but nonculturable pathogens in the food industry: current status and future prospects. Compr Rev Food Sci Food Saf 20:2146–2175
76. Lei S, Chen S, Zhong Q (2021) Digital PCR for accurate quantification of pathogens: principles, applications, challenges and future prospects. Int J Biol Macromol 184:750–759
77. Rane TD, Chen L, Zec HC, Wang T-H (2015) Microfluidic continuous flow digital loop-mediated isothermal amplification (LAMP). Lab Chip 15:776–782
78. Azizi M, Zaferani M, Cheong SH, Abbaspourrad A (2019) Pathogenic bacteria detection using RNA-based loop-mediated isothermal-amplification-assisted nucleic acid amplification via droplet microfluidics. ACS Sens 4:841–848. https://doi.org/10.1021/acssensors.8b01206
79. Wu H, Cao X, Meng Y, Richards D, Wu J, Ye Z, DeMello AJ (2022) DropCRISPR: a LAMP-Cas12a based digital method for ultrasensitive detection of nucleic acid. Biosens Bioelectron 211:114377
80. Zhang Z, Cheng Y, Li X, Chen L, Xu R, Qi X, Shao Y, Gao Z, Zhu M (2023) Bent-capillary-centrifugal-driven monodisperse droplet generator with its application for digital LAMP assay. Anal Chem 95:3028–3036. https://doi.org/10.1021/acs.analchem.2c05110
81. Schoepp NG, Schlappi TS, Curtis MS, Butkovich SS, Miller S, Humphries RM, Ismagilov RF (2017) Rapid pathogen-specific phenotypic antibiotic susceptibility testing using digital LAMP quantification in clinical samples. Sci Transl Med 9:eaal3693. https://doi.org/10.1126/scitranslmed.aal3693
82. Ma Y-D, Chang W-H, Wang C-H, Liu S-Y, Yen W-H, Luo K, You H-L, Wu J-J, Lee MS, Lee G-B (2016) A self-driven microfluidic chip through a rapid surface modification of PDMS and its application for digital loop-mediated amplification (LAMP). In: 2016 IEEE 11th annual international conference on nano/micro engineered and molecular systems (NEMS), pp 1–4. https://doi.org/10.1109/NEMS.2016.7758187
83. Rodriguez-Manzano J, Karymov MA, Begolo S, Selck DA, Zhukov DV, Jue E, Ismagilov RF (2016) Reading out single-molecule digital RNA and DNA isothermal amplification in nanoliter volumes with unmodified camera phones. ACS Nano 10:3102–3113. https://doi.org/10.1021/acsnano.5b07338
84. Hu Y, Xu P, Luo J, He H, Du W (2017) Absolute quantification of H5-subtype avian influenza viruses using droplet digital loop-mediated isothermal amplification. Anal Chem 89:745–750
85. Tan Y-L, Huang A-Q, Tang L-J, Jiang J-H (2021) Multiplexed droplet loop-mediated isothermal amplification with scorpion-shaped probes and fluorescence microscopic counting for digital quantification of virus RNAs. Chem Sci 12:8445–8451. https://doi.org/10.1039/D1SC00616A
86. Fan Z, Sun Y, Lin J-M (2022) Self-assembled inkjet printer for droplet digital loop-mediated isothermal amplification. Chemosensors 10:247. https://doi.org/10.3390/chemosensors10070247
87. Coelho BJ, Veigas B, Bettencourt L, Águas H, Fortunato E, Martins R, Baptista PV, Igreja R (2022) Digital microfluidics-powered real-time monitoring of isothermal DNA amplification of cancer biomarker. Biosensors 12:201. https://doi.org/10.3390/bios12040201
88. Konry T, Smolina I, Yarmush JM, Irimia D, Yarmush ML (2011) Ultrasensitive detection of low-abundance surface-marker protein using isothermal rolling circle amplification in a microfluidic nanoliter platform. Small 7:395–400. https://doi.org/10.1002/smll.201001620
89. Chi Y, Shi M, Wu Y, Wu Y, Chang Y, Liu M (2022) Single bacteria detection by droplet DNAzyme-coupled rolling circle amplification. Anal Methods 14:2244–2248. https://doi.org/10.1039/D2AY00656A
90. Kühnemund M, Hernández-Neuta I, Sharif MI, Cornaglia M, Gijs MAM, Nilsson M (2017) Sensitive and inexpensive digital DNA analysis by microfluidic enrichment of rolling circle amplified single-molecules. Nucleic Acids Res 45:e59. https://doi.org/10.1093/nar/gkw1324

91. Wang K, Bai X, Xue Y, Luo X, Dong J, Yang G, Ma K, Huang L, Zhou J, Wang J (2023) Absolute quantification of circRNA using digital reverse transcription-hyperbranched rolling circle amplification. Sensors Actuators B Chem 375:132893
92. Mazutis L, Araghi AF, Miller OJ, Baret J-C, Frenz L, Janoshazi A, Taly V, Miller BJ, Hutchison JB, Link D (2009) Droplet-based microfluidic systems for high-throughput single DNA molecule isothermal amplification and analysis. Anal Chem 81:4813–4821
93. Nishikawa Y, Hosokawa M, Maruyama T, Yamagishi K, Mori T, Takeyama H (2015) Monodisperse picoliter droplets for low-bias and contamination-free reactions in single-cell whole genome amplification. PLoS One 10:e0138733
94. Rhee M, Light YK, Meagher RJ, Singh AK (2016) Digital droplet multiple displacement amplification (ddMDA) for whole genome sequencing of limited DNA samples. PLoS One 11: e0153699. https://doi.org/10.1371/journal.pone.0153699
95. Sidore AM, Lan F, Lim SW, Abate AR (2016) Enhanced sequencing coverage with digital droplet multiple displacement amplification. Nucleic Acids Res 44:e66. https://doi.org/10.1093/nar/gkv1493
96. Shen F, Davydova EK, Du W, Kreutz JE, Piepenburg O, Ismagilov RF (2011) Digital isothermal quantification of nucleic acids via simultaneous chemical initiation of recombinase polymerase amplification reactions on SlipChip. Anal Chem 83:3533–3540
97. Tsaloglou M-N, Watson RJ, Rushworth CM, Zhao Y, Niu X, Sutton JM, Morgan H (2015) Real-time microfluidic recombinase polymerase amplification for the toxin B gene of Clostridium difficile on a SlipChip platform. Analyst 140:258–264. https://doi.org/10.1039/C4AN01683A
98. Schuler F, Schwemmer F, Trotter M, Wadle S, Zengerle R, von Stetten F, Paust N (2015) Centrifugal step emulsification applied for absolute quantification of nucleic acids by digital droplet RPA. Lab Chip 15:2759–2766. https://doi.org/10.1039/C5LC00291E
99. Kalsi S, Valiadi M, Tsaloglou M-N, Parry-Jones L, Jacobs A, Watson R, Turner C, Amos R, Hadwen B, Buse J, Brown C, Sutton M, Morgan H (2015) Rapid and sensitive detection of antibiotic resistance on a programmable digital microfluidic platform. Lab Chip 15:3065–3075. https://doi.org/10.1039/C5LC00462D
100. Genot AJ, Baccouche A, Sieskind R, Aubert-Kato N, Bredeche N, Bartolo JF, Taly V, Fujii T, Rondelez Y (2016) High-resolution mapping of bifurcations in nonlinear biochemical circuits. Nat Chem 8:760–767. https://doi.org/10.1038/nchem.2544
101. Okumura S, Gines G, Lobato-Dauzier N, Baccouche A, Deteix R, Fujii T, Rondelez Y, Genot AJ (2022) Nonlinear decision-making with enzymatic neural networks. Nature 610:496–501. https://doi.org/10.1038/s41586-022-05218-7
102. Jet T, Gines G, Rondelez Y, Taly V (2021) Advances in multiplexed techniques for the detection and quantification of microRNAs. Chem Soc Rev 50:4141–4161. https://doi.org/10.1039/D0CS00609B
103. Masurier A (2021) Molecular programming for the ultra-sensitive detection of biomarkers involved in neurodegenerative diseases. These de doctorat, Université Paris sciences et lettres
104. Ghadessy FJ, Ong JL, Holliger P (2001) Directed evolution of polymerase function by compartmentalized self-replication. Proc Natl Acad Sci 98:4552–4557. https://doi.org/10.1073/pnas.071052198
105. Milligan JN, Shroff R, Garry DJ, Ellington AD (2018) Evolution of a thermophilic strand-displacing polymerase using high-temperature isothermal compartmentalized self-replication. Biochemistry 57:4607–4619. https://doi.org/10.1021/acs.biochem.8b00200
106. Ellefson JW, Meyer AJ, Hughes RA, Cannon JR, Brodbelt JS, Ellington AD (2014) Directed evolution of genetic parts and circuits by compartmentalized partnered replication. Nat Biotechnol 32:97–101. https://doi.org/10.1038/nbt.2714
107. Dramé-Maigné A, Espada R, McCallum G, Sieskind R, Rondelez Y (2021) Directed evolution of enzymes based on in vitro programmable self-replication. bioRxiv:2021.04.22.440993. https://doi.org/10.1101/2021.04.22.440993

108. Lan F, Demaree B, Ahmed N, Abate AR (2017) Single-cell genome sequencing at ultra-high-throughput with microfluidic droplet barcoding. Nat Biotechnol 35:640–646. https://doi.org/10.1038/nbt.3880
109. Pellegrino M, Sciambi A, Treusch S, Durruthy-Durruthy R, Gokhale K, Jacob J, Chen TX, Geis JA, Oldham W, Matthews J, Kantarjian H, Futreal PA, Patel K, Jones KW, Takahashi K, Eastburn DJ (2018) High-throughput single-cell DNA sequencing of acute myeloid leukemia tumors with droplet microfluidics. Genome Res 28:1345–1352. https://doi.org/10.1101/gr.232272.117
110. Klein AM, Mazutis L, Akartuna I, Tallapragada N, Veres A, Li V, Peshkin L, Weitz DA, Kirschner MW (2015) Droplet barcoding for single-cell transcriptomics applied to embryonic stem cells. Cell 161:1187–1201. https://doi.org/10.1016/j.cell.2015.04.044
111. Zheng GXY, Terry JM, Belgrader P, Ryvkin P, Bent ZW, Wilson R, Ziraldo SB, Wheeler TD, McDermott GP, Zhu J, Gregory MT, Shuga J, Montesclaros L, Underwood JG, Masquelier DA, Nishimura SY, Schnall-Levin M, Wyatt PW, Hindson CM, Bharadwaj R, Wong A, Ness KD, Beppu LW, Deeg HJ, McFarland C, Loeb KR, Valente WJ, Ericson NG, Stevens EA, Radich JP, Mikkelsen TS, Hindson BJ, Bielas JH (2017) Massively parallel digital transcriptional profiling of single cells. Nat Commun 8:14049. https://doi.org/10.1038/ncomms14049
112. Kim SC, Clark IC, Shahi P, Abate AR (2018) Single-cell RT-PCR in microfluidic droplets with integrated chemical lysis. Anal Chem 90:1273–1279. https://doi.org/10.1021/acs.analchem.7b04050
113. Chung MT, Kurabayashi K, Cai D (2019) Single-cell RT-LAMP mRNA detection by integrated droplet sorting and merging. Lab Chip 19:2425–2434. https://doi.org/10.1039/C9LC00161A
114. Hyman LB, Christopher CR, Romero PA (2021) Single-cell nucleic acid profiling in droplets (SNAPD) enables high-throughput analysis of heterogeneous cell populations. Nucleic Acids Res 49:e103–e103. https://doi.org/10.1093/nar/gkab577
115. Pardi N, Hogan MJ, Porter FW, Weissman D (2018) mRNA vaccines – a new era in vaccinology. Nat Rev Drug Discov 17:261–279. https://doi.org/10.1038/nrd.2017.243
116. Bloom K, van den Berg F, Arbuthnot P (2021) Self-amplifying RNA vaccines for infectious diseases. Gene Ther 28:117–129. https://doi.org/10.1038/s41434-020-00204-y
117. Ellington AD, Szostak JW (1990) In vitro selection of RNA molecules that bind specific ligands. Nature 346:818–822
118. Guatelli JC, Whitfield KM, Kwoh DY, Barringer KJ, Richman DD, Gingeras TR (1990) Isothermal, in vitro amplification of nucleic acids by a multienzyme reaction modeled after retroviral replication. Proc Natl Acad Sci 87:1874–1878
119. Compton J (1991) Nucleic acid sequence-based amplification. Nature 350:91–92. https://doi.org/10.1038/350091a0
120. La Rocco MT, Wanger A, Ocera H, Macias E (1994) Evaluation of a commercial rRNA amplification assay for direct detection of Mycobacterium tuberculosis in processed sputum. Eur J Clin Microbiol Infect Dis 13:726–731. https://doi.org/10.1007/BF02276055
121. Ju Y, Yong Kim H, Ki Ahn J, Gyu Park H (2021) Ultrasensitive version of nucleic acid sequence-based amplification (NASBA) utilizing a nicking and extension chain reaction system. Nanoscale 13:10785–10791. https://doi.org/10.1039/D1NR00564B
122. Ju Y, Song Y, Kim J, Kim H, Li Y, Lee KG, Lee SJ, Park HG (2023) A novel isothermal digital amplification system and its application for absolute quantification of respiratory infectious virus. Biosens Bioelectron X 13:100317. https://doi.org/10.1016/j.biosx.2023.100317
123. Tucker BJ, Breaker RR (2005) Riboswitches as versatile gene control elements. Curr Opin Struct Biol 15:342–348
124. Park SV, Yang J-S, Jo H, Kang B, Oh SS, Jung GY (2019) Catalytic RNA, ribozyme, and its applications in synthetic biology. Biotechnol Adv 37:107452
125. George L, Indig FE, Abdelmohsen K, Gorospe M (2018) Intracellular RNA-tracking methods. R Soc Open Biol 8:180104

126. Ali MH, Elsherbiny ME, Emara M (2019) Updates on aptamer research. Int J Mol Sci 20:2511
127. Ryckelynck M, Baudrey S, Rick C, Marin A, Coldren F, Westhof E, Griffiths AD (2015) Using droplet-based microfluidics to improve the catalytic properties of RNA under multiple-turnover conditions. RNA 21:458–469
128. Autour A, Westhof E, Ryckelynck M (2016) ISpinach: a fluorogenic RNA aptamer optimized for in vitro applications. Nucleic Acids Res 44:2491–2500. https://doi.org/10.1093/nar/gkw083
129. Autour A, Bouhedda F, Cubi R, Ryckelynck M (2019) Optimization of fluorogenic RNA-based biosensors using droplet-based microfluidic ultrahigh-throughput screening. Methods 161:46–53. https://doi.org/10.1016/j.ymeth.2019.03.015
130. Autour A, Jeng SCY, Cawte AD, Abdolahzadeh A, Galli A, Panchapakesan SSS, Rueda D, Ryckelynck M, Unrau PJ (2018) Fluorogenic RNA Mango aptamers for imaging small non-coding RNAs in mammalian cells. Nat Commun 9:1–12
131. Trachman RJ, Autour A, Jeng SCY, Abdolahzadeh A, Andreoni A, Cojocaru R, Garipov R, Dolgosheina EV, Knutson JR, Ryckelynck M (2019) Structure and functional reselection of the Mango-III fluorogenic RNA aptamer. Nat Chem Biol 15:472–479
132. Bouhedda F, Fam KT, Collot M, Autour A, Marzi S, Klymchenko A, Ryckelynck M (2020) A dimerization-based fluorogenic dye-aptamer module for RNA imaging in live cells. Nat Chem Biol 16(69):76. https://doi.org/10.1038/s41589-019-0381-8
133. Aufinger L, Simmel FC (2018) Artificial gel-based organelles for spatial organization of cell-free gene expression reactions. Angew Chem Int Ed 57:17245–17248. https://doi.org/10.1002/anie.201809374
134. Weitz M, Kim J, Kapsner K, Winfree E, Franco E, Simmel FC (2014) Diversity in the dynamical behaviour of a compartmentalized programmable biochemical oscillator. Nat Chem 6:295–302. https://doi.org/10.1038/nchem.1869
135. Schwarz-Schilling M, Kim J, Cuba C, Weitz M, Franco E, Simmel FC (2016) Building a synthetic transcriptional oscillator. Cell Cycle Oscil Methods Protoc:185–199
136. Ayukawa S, Takinoue M, Kiga D (2011) RTRACS: a modularized RNA-dependent RNA transcription system with high programmability. Acc Chem Res 44:1369–1379. https://doi.org/10.1021/ar200128b
137. Mizuuchi R, Ichihashi N (2018) Sustainable replication and coevolution of cooperative RNAs in an artificial cell-like system. Nat Ecol Evol 2:1654–1660. https://doi.org/10.1038/s41559-018-0650-z
138. Furubayashi T, Ueda K, Bansho Y, Motooka D, Nakamura S, Mizuuchi R, Ichihashi N (2020) Emergence and diversification of a host-parasite RNA ecosystem through Darwinian evolution. elife 9:e56038. https://doi.org/10.7554/eLife.56038
139. Mizuuchi R, Furubayashi T, Ichihashi N (2022) Evolutionary transition from a single RNA replicator to a multiple replicator network. Nat Commun 13:1460. https://doi.org/10.1038/s41467-022-29113-x
140. Kwon Y-C, Jewett MC (2015) High-throughput preparation methods of crude extract for robust cell-free protein synthesis. Sci Rep 5:8663
141. Cole SD, Beabout K, Turner KB, Smith ZK, Funk VL, Harbaugh SV, Liem AT, Roth PA, Geier BA, Emanuel PA (2019) Quantification of interlaboratory cell-free protein synthesis variability. ACS Synth Biol 8:2080–2091
142. Foshag D, Henrich E, Hiller E, Schäfer M, Kerger C, Burger-Kentischer A, Diaz-Moreno I, García-Mauriño SM, Dötsch V, Rupp S (2018) The E. coli S30 lysate proteome: a prototype for cell-free protein production. New Biotechnol 40:245–260
143. Garenne D, Beisel CL, Noireaux V (2019) Characterization of the all-E. coli transcription-translation system myTXTL by mass spectrometry. Rapid Commun Mass Spectrom 33:1036–1048
144. Kung H-F, Redfield B, Treadwell B, Eskin B, Spears C, Weissbach H (1977) DNA-directed in vitro synthesis of beta-galactosidase. Studies with purified factors. J Biol Chem 252:6889–6894

145. Doerr A, De Reus E, Van Nies P, Van der Haar M, Wei K, Kattan J, Wahl A, Danelon C (2019) Modelling cell-free RNA and protein synthesis with minimal systems. Phys Biol 16: 025001
146. Lavickova B, Maerkl SJ (2019) A simple, robust, and low-cost method to produce the PURE cell-free system. ACS Synth Biol 8:455–462
147. Grasemann L, Lavickova B, Elizondo-Cantú MC, Maerkl SJ (2021) OnePot PURE cell-free system. JoVE J Vis Exp:e62625
148. Tawfik DS, Griffiths AD (1998) Man-made cell-like compartments for molecular evolution. Nat Biotechnol 16:652
149. Bernath K, Magdassi S, Tawfik DS (2005) Directed evolution of protein inhibitors of DNA-nucleases by in vitro compartmentalization (IVC) and nano-droplet delivery. J Mol Biol 345:1015–1026
150. Mastrobattista E (2005) Discovering novel evolutionary pathways to b-galactosidases using in vitro compartmentalization and fluorescence activated sorting of double emulsions. Chem Biol 12:1291–1300
151. Lu W, Levy M, Kincaid R, Ellington AD (2014) Directed evolution of the substrate specificity of biotin ligase. Biotechnol Bioeng 111:1071–1081
152. Prodanović R, Lloyd Ung W, Đurđić KI, Fischer R, Weitz DA, Ostafe R (2020) A high throughput screening system based on droplet microfluidics for glucose oxidase gene libraries. Molecules 25:1–12. https://doi.org/10.3390/molecules25102418
153. van Loo B, Heberlein M, Mair P, Zinchenko A, Schuurmann J, Eenink BDG, Holstein JM, Dilkaute C, Jose J, Hollfelder F (2019) High-throughput, lysis-free screening for sulfatase activity using Escherichia coli autodisplay in microdroplets. ACS Synth Biol 8:2690–2700
154. Fallah-Araghi A, Baret J-C, Ryckelynck M, Griffiths AD (2012) A completely in vitro ultrahigh-throughput droplet-based microfluidic screening system for protein engineering and directed evolution. Lab Chip 12:882–891
155. Sakatani Y, Mizuuchi R, Ichihashi N (2019) In vitro evolution of phi29 DNA polymerases through compartmentalized gene expression and rolling-circle replication. Protein Eng Des Sel 32:481–487. https://doi.org/10.1093/protein/gzaa011
156. Bertschinger J, Neri D (2004) Covalent DNA display as a novel tool for directed evolution of proteins in vitro. Protein Eng Des Sel 17:699–707
157. Houlihan G, Gatti-Lafranconi P, Kaltenbach M, Lowe D, Hollfelder F (2014) An experimental framework for improved selection of binding proteins using SNAP display. J Immunol Methods 405:47–56
158. Diamante L, Gatti-Lafranconi P, Schaerli Y, Hollfelder F (2013) In vitro affinity screening of protein and peptide binders by megavalent bead surface display. Protein Eng Des Sel 26:713–724
159. Sieskind R (2019) In vitro enzyme directed evolution using DNA artificial networks
160. Fryer T (2022) Polyacrylamide hydrogel bead display for ultra-high throughput screening and evolution of therapeutic proteins
161. Fryer T, Rogers JD, Mellor C, Kohler TN, Minter R, Hollfelder F (2022) Gigavalent display of proteins on monodisperse polyacrylamide hydrogels as a versatile modular platform for functional assays and protein engineering. ACS Cent Sci 8:1182–1195
162. Bashaw BT, Peterson AT, Schildknecht KT, Tauchen AT, Tittor RT, Gomez P, McKlin S, Chen Q, Mazariegos C, Zhou J (2020) A microfluidics approach to the high-throughput generation of quantifiable cell-free protein synthesis systems. FASEB J 34:1–1
163. Sierra AMR, Arold ST, Grünberg R (2022) Efficient multi-gene expression in cell-free droplet microreactors. PLoS One 17:e0260420. https://doi.org/10.1371/journal.pone.0260420
164. Hori Y, Kantak C, Murray R, Abate A (2017) Cell-free extract based optimization of biomolecular circuits with droplet microfluidics. Lab Chip 17:3037–3042. https://doi.org/10.1039/C7LC00552K
165. Charon J, Manteca A, Innis CA (2019) Using the bacterial ribosome as a discovery platform for peptide-based antibiotics. Biochemistry 58. https://doi.org/10.1021/acs.biochem.8b00927

166. Mookherjee N, Anderson MA, Haagsman HP, Davidson DJ (2020) Antimicrobial host defence peptides: functions and clinical potential. Nat Rev Drug Discov 19:311–332
167. Nuti N, Rottmann P, Stucki A, Koch P, Panke S, Dittrich PS (2022) A multiplexed cell-free assay to screen for antimicrobial peptides in double emulsion droplets. Angew Chem Int Ed 61:e202114632
168. Sun H, Hu N, Wang J (2022) Application of microfluidic technology in antibody screening. Biotechnol J 17:2100623
169. Jacková B, Mottet G, Rudiuk S, Morel M, Baigl D (2022) DNA-encoded immunoassay in picoliter drops: a minimal cell-free approach. bioRxiv. 2022–09
170. Ding R, Hung K-C, Mitra A, Ung LW, Lightwood D, Tu R, Starkie D, Cai L, Mazutis L, Chong S (2020) Rapid isolation of antigen-specific B-cells using droplet microfluidics. RSC Adv 10:27006–27013
171. Jewett MC, Forster AC (2010) Update on designing and building minimal cells. Curr Opin Biotechnol 21:697–703
172. Jia H, Heymann M, Bernhard F, Schwille P, Kai L (2017) Cell-free protein synthesis in micro compartments: building a minimal cell from biobricks. New Biotechnol 39:199–205. https://doi.org/10.1016/j.nbt.2017.06.014
173. Laohakunakorn N, Grasemann L, Lavickova B, Michielin G, Shahein A, Swank Z, Maerkl SJ (2020) Bottom-up construction of complex biomolecular systems with cell-free synthetic biology. Front Bioeng Biotechnol 8
174. Godino E, López JN, Foschepoth D, Cleij C, Doerr A, Castellà CF, Danelon C (2019) De novo synthesized min proteins drive oscillatory liposome deformation and regulate FtsA-FtsZ cytoskeletal patterns. Nat Commun 10:4969. https://doi.org/10.1038/s41467-019-12932-w
175. Postek W, Garstecki P (2022) Droplet microfluidics for high-throughput analysis of antibiotic susceptibility in bacterial cells and populations. Acc Chem Res 55:605–615
176. Findlay HE, Harris NJ, Booth PJ (2016) In vitro synthesis of a major facilitator transporter for specific active transport across droplet interface bilayers. Sci Rep 6:39349
177. Tang TD, Cecchi D, Fracasso G, Accardi D, Coutable-Pennarun A, Mansy SS, Perriman AW, Anderson JR, Mann S (2018) Gene-mediated chemical communication in synthetic protocell communities. ACS Synth Biol 7:339–346
178. Adamala KP, Martin-Alarcon DA, Guthrie-Honea KR, Boyden ES (2017) Engineering genetic circuit interactions within and between synthetic minimal cells. Nat Chem 9:431–439
179. Gruner P, Riechers B, Semin B, Lim J, Johnston A, Short K, Baret J-C (2016) Controlling molecular transport in minimal emulsions. Nat Commun 7:1–9
180. Booth MJ, Schild VR, Graham AD, Olof SN, Bayley H (2016) Light-activated communication in synthetic tissues. Sci Adv 2:e1600056
181. Dupin A, Simmel FC (2019) Signalling and differentiation in emulsion-based multi-compartmentalized in vitro gene circuits. Nat Chem 11:32–39. https://doi.org/10.1038/s41557-018-0174-9
182. Dupin A, Aufinger L, Styazhkin I, Rothfischer F, Kaufmann BK, Schwarz S, Galensowske N, Clausen-Schaumann H, Simmel FC (2022) Synthetic cell–based materials extract positional information from morphogen gradients. Sci Adv 8:eabl9228. https://doi.org/10.1126/sciadv.abl9228
183. van Nies P, Westerlaken I, Blanken D, Salas M, Mencía M, Danelon C (2018) Self-replication of DNA by its encoded proteins in liposome-based synthetic cells. Nat Commun 9:1583. https://doi.org/10.1038/s41467-018-03926-1
184. Okauchi H, Sakatani Y, Otsuka K, Ichihashi N (2020) Minimization of elements for isothermal DNA replication by an evolutionary approach. ACS Synth Biol 9:1771–1780. https://doi.org/10.1021/acssynbio.0c00137
185. Okauchi H, Ichihashi N (2021) Continuous cell-free replication and evolution of artificial genomic DNA in a compartmentalized gene expression system. ACS Synth Biol 10:3507–3517. https://doi.org/10.1021/acssynbio.1c00430

186. Miyachi R, Shimizu Y, Ichihashi N (2022) Transfer RNA synthesis-coupled translation and DNA replication in a reconstituted transcription/translation system. ACS Synth Biol 11:2791–2799. https://doi.org/10.1021/acssynbio.2c00163
187. Ueno H, Sawada H, Soga N, Sano M, Nara S, Tabata KV, Su'etsugu M, Noji H (2021) Amplification of over 100 kbp DNA from single template molecules in femtoliter droplets. ACS Synth Biol 10:2179–2186. https://doi.org/10.1021/acssynbio.0c00584
188. Floyd BM, Marcotte EM (2022) Protein sequencing, one molecule at a time. Annu Rev Biophys 51:181–200
189. Mair P, Gielen F, Hollfelder F (2017) Exploring sequence space in search of functional enzymes using microfluidic droplets. Curr Opin Chem Biol 37:137–144
190. Conchouso D, Al-Ma'abadi A, Behzad H, Alarawi M, Hosokawa M, Nishikawa Y, Takeyama H, Mineta K, Gojobori T (2021) Integration of droplet microfluidic tools for single-cell functional metagenomics: an engineering head start. Genomics Proteomics Bioinformatics 19:504–518
191. Schwille P, Spatz J, Landfester K, Bodenschatz E, Herminghaus S, Sourjik V, Erb TJ, Bastiaens P, Lipowsky R, Hyman A (2018) MaxSynBio: avenues towards creating cells from the bottom up. Angew Chem Int Ed 57:13382–13392
192. Frischmon C, Sorenson C, Winikoff M, Adamala KP (2021) Build-a-cell: engineering a synthetic cell community. Life 11:1176

Adv Biochem Eng Biotechnol (2023) 185: 129–146
https://doi.org/10.1007/10_2023_225

Published online: 13 June 2023

eCell Technology for Cell-Free Protein Synthesis, Biosensing, and Remediation

Damian Van Raad and Thomas Huber

Contents

Abstract The eCell technology is a recently introduced, specialized protein production platform with uses in a multitude of biotechnological applications. This chapter summarizes the use of eCell technology in four selected application areas. Firstly, for detecting heavy metal ions, specifically mercury, in an in vitro protein expression system. Results show improved sensitivity and lower limit of detection compared to comparable in vivo systems. Secondly, eCells are semipermeable, stable, and can be stored for extended periods of time, making them a portable and accessible technology for bioremediation of toxicants in extreme environments. Thirdly and fourthly, applications of eCell technology are shown to facilitate expression of correctly folded disulfide-rich proteins and incorporate chemically interesting derivatives of amino acids into proteins which are toxic to in vivo protein expression. Overall, eCell technology presents a cost-effective and efficient method for biosensing, bioremediation, and protein production.

D. Van Raad and T. Huber (✉)
Research School of Chemistry, Australian National University, Canberra, ACT, Australia
e-mail: damian.vanraad@anu.edu.au; t.huber@anu.edu.au

Graphical Abstract

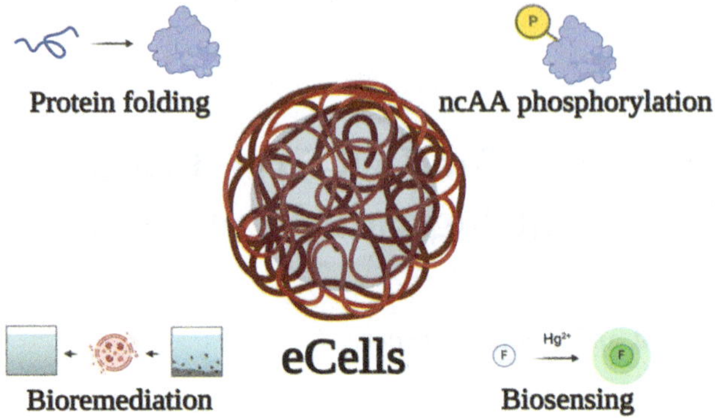

Abbreviations

AA	Amino acid
CFPS	Cell-free protein synthesis
E. coli	*Escherichia coli*
GSH	Glutathione reduced
GSSG	Glutathione oxidized
LbL	Layer-by-layer assembly
ncAA	Non-canonical amino acid
O-phospho-L-serine, SeP	O-phospho-L-threonine, pThr
PylRS	Pyrrolysl-tRNA synthetase
S30	Supernatant 30,000 g

1 Introduction

CFPS systems are characterized by the functional use of transcription/translation macromolecules for recombinant protein expression without the use of living cells [1]. In the preparatory process, cell walls are removed from living cells and cytosolic components condensed into a soup of translationally active lysate [2, 3]. Lysates are defined as either S12 or S30 extracts, which is termed through its preparation process. Cells are harvested at the middle of the exponential phase of low-density growth and then processed using a pressure cell homogenizer or sonication [4]. The supernatant is then centrifuged at 30,000 g or 12,000 g to remove cell wall fragments, genomic DNA, and mRNA, hence the name Supernatant 30,000 g/Supernatant 12,000 g (S30 or S12) [5]. This lysate holds all the necessary macromolecules that facilitate energy production, transcription, translation, and protein folding in the cell [6]. The addition of reagents (energy substrates, cofactors, and salts), the plasmid or PCR product of interest and amino acids (AAs) allows production of recombinant protein [7]. Through the removal of the cell wall, CFPS lysates lose the property of strictly defined cell homeostasis and become an open system amenable to chemical modification with the use of reductive, oxidative, or toxic reagents [8, 9].

The eCell system is a novel approach to in vitro cell-free protein synthesis that employs the principles of Layer-by-Layer (LbL) polymer assembly to encapsulate bacteria [10, 11]. The encapsulation of bacteria circumvents the need for lysate preparation, while still retaining the small cell-like structure in in vitro reactions. This results in a cost-effective and easily accessible method for cell-free protein synthesis. The semipermeable capsule surrounding the bacteria is formed by the sequential deposition of charged polyelectrolytes onto the cell surface, which creates a rigid shell that retains all components of the bacteria. The native cell wall of the bacteria is lysed using a hydrolytic enzyme, leaving a layered and semipermeable capsule surrounding the cell's constituents. The eCell system offers several advantages, including versatility and accessibility, as it can be used with conventional and non-conventional energy generation systems. The plasmid is already ready for translation and requires only the CFPS buffer to induce expression, making it a highly attractive platform for cell-free protein synthesis. eCells are non-living entities that maintain many of the advantages of living cells while allowing full chemical control of the environment. By encapsulating cells, the risk of secondary contamination is eliminated, and the efficiency of decontamination can be increased by using enzymes specific to the contaminants. Additionally, enzymes encapsulated within eCells demonstrate increased stability and a longer shelf life compared to isolated or purified enzymes used for bioremediation. By overexpressing bioremediation enzymes in eCells, an in vitro decontamination system can be applied to address toxicant and contamination issues such as pesticides and heavy metal detection.

In this chapter, we utilize the eCell platform in a diverse array of applications (Fig. 1). The eCell platform has been shown to enable the detection of heavy metal ions at sub-ppb levels, using a published biosensor for Hg^{2+} ions [18]. The detection is straightforward, regulated, and highly stringent. For bioremediation, an engineered phosphotriesterase (PTE), previously reported by Campbell et al. [23], is employed for organophosphate decontamination in a range of extreme environments. The utilization of eCells in saline environments serves as a demonstration of the robustness of the eCell technology. Additionally, acidic and basic environments pose significant challenges to bioremediation strategies as they can negatively impact the viability of the organism. The protective capsule of the eCell allows for the localization of enzymes, resulting in a stabilization effect and increased efficiency of decontamination in such environments.

The eCell platform has been demonstrated to be a valuable addition to the toolkit of synthetic biology, particularly in the production of conformationally correct disulfide-bonded proteins. Using eCell CFPS in combination with protein disulfide isomerase (PDI) and an altered redox potential of the endoplasmic reticulum (−208 mV), it is possible to achieve efficient and accurate folding of difficult to

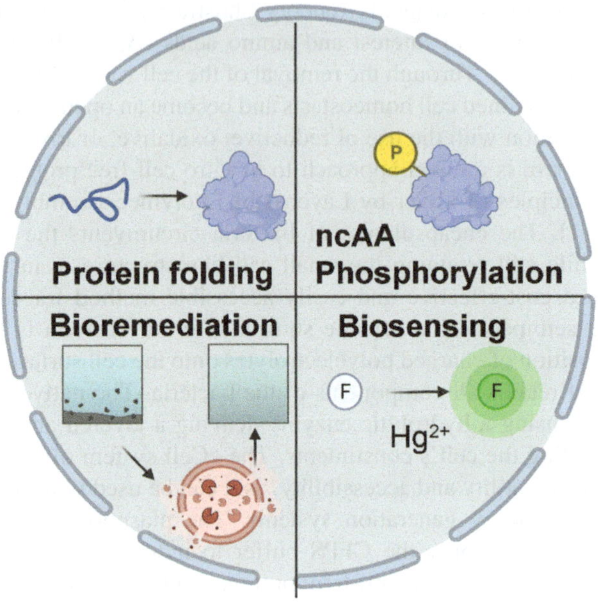

Fig. 1 The eCell system has applications in the fields of biosensing, bioremediation, production of proteins with non-canonical amino acids (ncAA), and protein folding. In biosensing, the semipermeable capsule of the eCell system can be used as a stable platform for the detection of analytes. In bioremediation, the eCell system can be used to encapsulate and protect proteins related to the biodegradation of pollutants, allowing for targeted and efficient removal of contaminants. The eCell system can be used to enhance correct protein folding with protein disulfide isomerase (PDI), a crucial enzyme in protein folding. The semipermeable capsule of the eCell system allows for the retention of both the protein and PDI, creating a controlled environment to produce disulfide-bonded proteins. In addition, eCells have been shown to be successful in incorporation of phosphorylated ncAAs

produce disulfide-rich proteins. As demonstrated, this method allows for soluble expression of neuritin, a human neurotrophic factor. Additionally, active human light chain enterokinase was produced and was found to efficiently cleave its protein target after its DDDDK recognition site. These results highlight the versatility and utility of eCells in the field of synthetic biology. The utilization of eCell technology allows for more advanced modifications of proteins, such as the incorporation of site-specific phosphorylation and/or fluorination of residues. The implementation of a previously published system for the incorporation of pThr and SeP into eCells does not only reduce reagent costs but also demonstrates the functionalization of bacterial cell lines into eCells using an auxiliary plasmid. Additionally, utilizing eCell technology may address issues of reproducibility in the incorporation of pThr using in vivo and standard CFPS methods.

2 Biosensing

CFPS systems have been applied in biosensing of a wide range of analytes [12–14]. CFPS has made biosensing more portable, through utilizing on the go fluorescence/absorbance output and complex CFPS paper detection using freeze dried reagents [15, 16]. In-cell based biosensors have been used to demonstrate the detection of contaminants in natural environments and the sensitive detection of heavy metal ions [17, 18]. Biosensors for mercury ions have been developed through engineering of the mercury responsive MerR reporter system [19, 20]. MerR responds to Hg^{2+} ions in a repressor-activator for its cognate promoter P_{merT}, which controls expression of a green-fluorescent protein and allows output characterization with Hg^{2+} concentrations to be detected at sub-ppb levels [20]. Despite the advantages of in vivo bioremediation and biosensing, in situ use of GMOs presents significant biosafety concerns that prevent large-scale use of living organisms in the natural environment [21]. Cell-free synthetic biology in conjunction with biosensing offers a promising mechanism for circumventing secondary contamination through GMO release [22]. In this way, biosensors can be safely used to detect toxic and pathogenic contaminants in pristine natural environments.

To demonstrate the eCell technology for biosensing, a plasmid pSB3K3 which constitutively expresses the repressor MerR, which binds to Hg^{2+} ions and induces the transcription of P_{merT} for green-fluorescent protein (GFP) expression (addgene plasmid #123148) was transformed in Xjb(DE3)* cell line for eCell CFPS and in vivo detection of mercury [18]. Expression of GFP was detected on a plate reader to compare Hg^{2+} detection and sensitivity between in vivo and in vitro systems. Ranges of Hg^{2+} concentrations from 0.01 to 1 μM were used in the comparison of eCell with in vivo culture biosensing (Fig. 2).

Sensitive detection of Hg^{2+} ions is accomplished in the eCell in vitro expression system. Fluorescence-based analysis revealed a significant increase in GFP expression in eCells exposed to increasing concentrations of Hg^{2+}, indicating a high level of sensitivity in detecting the presence of the heavy metal ions. Comparison of the in vitro system to in vivo expression revealed a significant increase in sensitivity in the eCells, which is likely due to the higher permeability of the artificial cells to Hg^{2+} ions compared to living cells. The semipermeable nature of the eCells allows for passive diffusion of the metal ions, as opposed to relying on controlled cellular uptake, resulting in a more stringent detection limit. The limit of detection for Hg^{2+} using eCells was found to be 0.01 μM, which is approximately 10 times better than the limit of detection observed in in vivo expression (0.1 μM). The results obtained were consistent with previously published data using similar constructs for Hg^{2+} detection in vivo, further validating the utility of eCells as a sensitive and reliable biosensing system. The eCells have been demonstrated to be an efficient and cost-effective biosensing tool, as they can be stored for extended periods of time without loss of protein expression capability and can be used as a portable biosensing system. Furthermore, the utilization of eCells in natural resource analysis poses minimal risk to the environment as they are non-living entities.

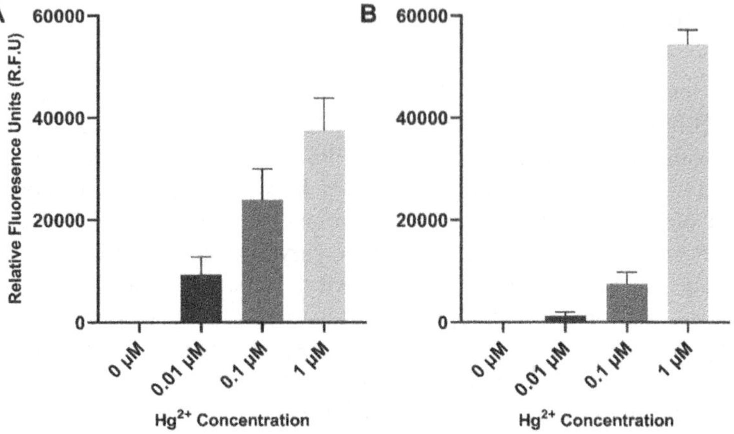

Fig. 2 Sensor output from 0 μM, 0.01 μM, 0.1 μM, and 1 μM of Hg^{2+} ions using GFP as a fluorescent reporter. (**a**) Fluorescent readout of GFP expressed in 3 mL of CFPS buffer using encapsulated and lysed eCells. (**b**) Fluorescent readout of GFP expressed in 25 mL in vivo expression of GFP. Fluorescence was normalized by weighing 50 mg of wet cell-mass in 3 mL to 50 mg of eCells in 3 mL and measuring 100 μL from triplicate expressions. Error bars represent standard deviation of three replicates

3 Bioremediation

Bioremediation is the functional use of either microbes or reconstituted enzymes to remove toxic molecules from the natural environment [23]. This has been used efficiently to transform toxicants into unreactive molecules, mitigating damage to natural resources [22–24]. The effectiveness of decontamination is linked to the efficiency of the enzyme. Genetic engineering has increased the efficiency of bioremediation enzymes, which is typically done by increasing the enzyme's catalytic efficiency. An example of this was through a directed evolution of an engineered phosphotriesterase (PTE) metalloenzyme originating from *Pseudomonas diminuta* [25]. This enzyme, which has been found in organisms living in organophosphate (parathion) contaminated soil, has been subjected to directed evolution to increase its catalytic specificity for phosphotriesters.

We hypothesized that eCells would be particularly suitable for bioremediation of toxicants in extreme environments. It was tested through the examination of their ability to inactivate the organophosphate pesticide derivative, paraoxon-ethyl. eCells were engineered to express an active phosphotriesterase (PTE) enzyme, which facilitates the hydrolysis of paraoxon-ethyl into diethylphosphoric acid and p-nitrophenol. The latter compound absorbs light at a wavelength of 405 nm, which was monitored by photometric measurements. The experiment was conducted by encapsulating PTE-overexpressed Xjb(DE3)* cells, lysing their cell wall, and then quantifying their ability to hydrolyze paraoxon-ethyl by measuring absorbance at 30 s intervals for 20 min. A total of 0.1 mg of eCells with overexpressed PTE and

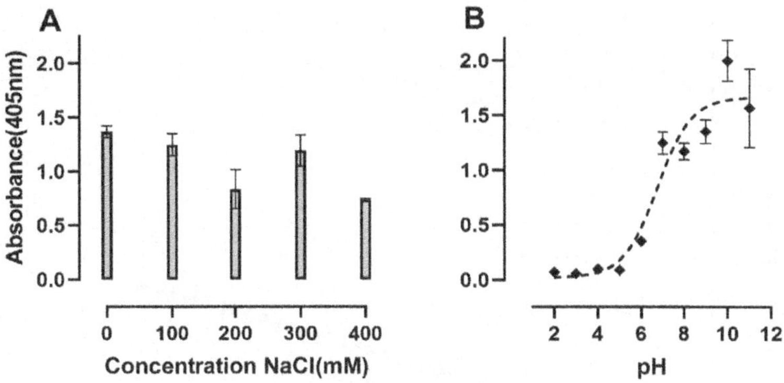

Fig. 3 (**a**) The impact of salinity on phosphotriesterase (PTE) activity was investigated by examining different salinity levels at 20-min endpoints. Salinity levels ranged from 100 to 400 mM NaCl. Results show that PTE activity is minimally affected by salinity up to a salt concentration of 300 mM, but decreases at slightly at 400 mM. (**b**) The influence of pH on PTE activity was evaluated by analyzing the final time point of each reaction across a pH range of 2 to 11. Results indicate no detectable activity at acidic pH, while PTE activity remains consistently active at higher pH levels

200 μl of reaction buffer (20 mM Tris–HCl, 100 mM NaCl, 100 μM $ZnCl_2$, and 100 μM ethyl paraoxon) were used in the assay. The results of this experiment demonstrate the potential of eCells as a viable bioremediation tool for toxicants in extreme environments, specifically for the inactivation of organophosphate compounds.

Figure 3 indicates that eCells containing PTE can effectively bioremediate paraoxon-ethyl, a toxic organophosphate pesticide derivative, in extreme environments, such as high salinity and pH. We observed robust PTE activity in eCells even at 400 mM (2.3% w/w) sodium chloride concentration, which approximates the salinity of sea water. The activity of the PTE enzyme was only slightly reduced in these conditions, retaining approximately 70% activity of the control. Additionally, we observed that PTE activity in eCells displayed a robustness to environmental pH changes, with increased activity at higher pH. However, acidic pH had a significant impact on the activity of encapsulated PTE, resulting in a significant decrease when exposed to pH less than 7. This loss of function at low pH is likely due to the compromised stability of the enzyme rather than changes in the properties of the eCells. On the other hand, PTE in eCells were able to withstand the effects of very basic pH with no effect on activity. The results suggest that eCells remove the need for purification of enzymes and allow their release in a wide variety of environments without secondary contamination. This is important as current bioremediation strategies utilize living microbes to decontaminate excavated material. These findings demonstrate the potential utility of eCells as a bioremediation tool for real-world environments.

An important feature of eCells is that they can be lyophilized and retain their function after re-hydration. This was demonstrated by encapsulating and lysing

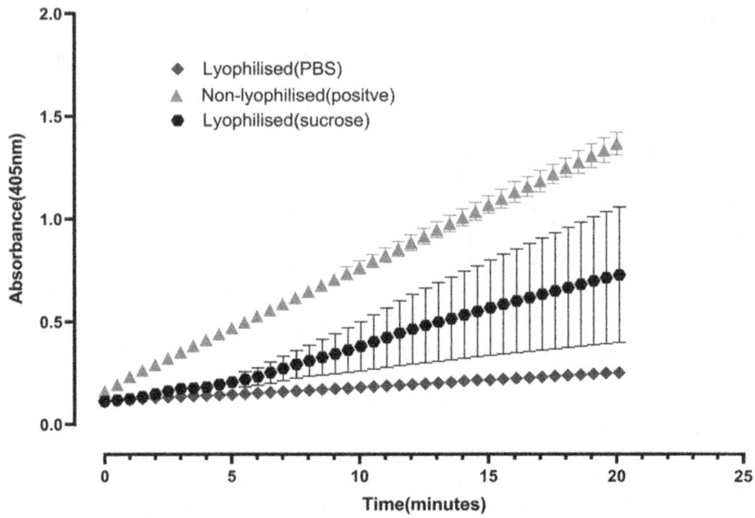

Fig. 4 Time-dependent analysis of the hydrolysis of paraoxon using encapsulated cells that have been genetically modified to overexpress phosphotriesterase (PTE). The enzymatic activity of PTE was evaluated under lyophilization conditions. The experimental groups included a positive control of encapsulated cells with overexpressed PTE, lyophilized eCells with overexpressed PTE and lyoprotecting groups (sucrose), and eCells with overexpressed PTE that were lyophilized from phosphate-buffered saline (PBS) buffer. The results show that the introduction of lyoprotecting groups, such as sucrose, led to enhanced stabilization of overexpressed PTE, as demonstrated by the observed difference between lyophilization with and without sucrose

overexpressed PTE Xjb(DE3)* eCells and lyophilizing them using two different buffers: phosphate-buffered saline (PBS) and sucrose (0.1 g/L). Subsequently, 0.1 mg of the lyophilized eCells were incubated with a reaction buffer and absorbance measurements were taken at regular intervals over a 20-min period.

The findings of the lyophilization experiments demonstrate that eCells containing PTE exhibit considerable enzymatic activity even after lyophilization and storage in both PBS and sucrose buffer (Fig. 4). The observed decline in PTE activity in eCells stored in PBS buffer as compared to those in sucrose buffer suggests that sucrose serves as an effective lyoprotectant for eCells. These outcomes validate the potential of eCell-based preservation methods as an alternative to traditional lyophilization of purified protein for the extended storage of PTE enzymatic activity. Additionally, the confinement of PTE within eCells is advantageous, preventing dispersion of the enzyme in larger aqueous environments. Nonetheless, the lyophilization process can have an impact on the enzyme activity and optimizing the lyophilization conditions such as temperature, time, and buffer composition is critical for maintaining enzyme stability.

Like in biosensing, the use of eCells for bioremediation is simple, inexpensive, and does not require the purification of enzymes to inactivate toxicants. The stabilizing effect of eCells allows for the use and storage of proteins with greater protection than reconstituted enzymes. This may be due to the crowded environment of eCells, which act as compartmentalized single cells, rather than the dispersion of reconstituted enzymes. As a result, enzymes within capsules are more stable and have a longer shelf life compared to isolated or purified enzymes used for bioremediation purposes. The activity of PTE-overexpressed eCells is reliable in variable environments, with activity retained even when exposed to alkaline pH or high salinity (Fig. 3a, b). Bioremediation using living microbes would often not be possible in these extreme environments. eCells can be applied to contamination issues such as pesticides, eliminating the need for microbial decontamination of excavated material. This can be achieved by overexpressing a specific enzyme to be applied to large amounts of contaminated excavated material or aqueous natural resources. The portability of eCells is enhanced by its ability to be lyophilized and retain its bioremediation capacity. A loss of 50% activity is observed during the lyophilization process when using sucrose as a lyoprotectant (Fig. 4). However, the retention of activity in a dry capsule form is advantageous for the portability of eCells.

4 Functional Expression of Disulfide-Rich Proteins

Preparation of disulfide-rich proteins using cell-based expression systems still suffers from several issues. Typically, these proteins are hard to express in bacterial cells and/or form inclusion bodies which require refolding using denaturants [26]. Refolding proteins is laborious as it requires gradual reduction of high concentrations of a denaturant (i.e., urea, guanidium, arginine) and is a slow process [27]. In addition, the process of folding a protein from inclusion bodies and restoring its activity can be inconsistent when denaturants are used. An elegant way to improve the yield of correctly folded disulfide-rich proteins is through modification of the reductive potential of the CFPS lysate and the additional use of folding chaperones. A homodimeric therapeutic biomolecule, porcine Interleukin-12 (IL-12) composed of two subunits of p40, was successfully refolded using a ratio of reduced/oxidized glutathione (GSH/GSSG), protein disulfide isomerase (PDI) from *Homo sapiens* and other chaperones [28]. After p40 overexpression in *E. coli* and solubilization in urea, PDI using a ratio of GSH/GSSG was found to catalyze the formation and breakage of disulfide bonds into the conformationally correct structure. This PDI mediated redox renaturation was found to be the superior method for correct folding over spontaneous refolding with urea.

The use of eCells in biotechnology offers several advantages, including stability and control over the expression environment. By overexpressing protein disulfide isomerase (PDI) in eCells and altering the expression conditions to mimic the redox potential of the endoplasmic reticulum, it is possible to produce Disulfide-rich proteins in a soluble, active form. This was demonstrated using a maltose binding protein fused to a DDDDK linker and a human light chain enterokinase (MBP-DDDDK-EK) with a 6x polyhistidine tag. The pET vector with ampicillin

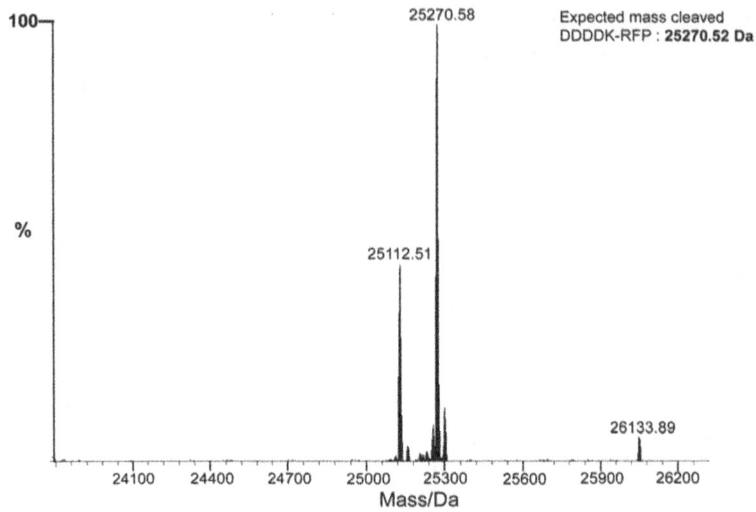

Fig. 5 High-resolution MS of RFP modified with N-terminal 6xHis tag and DDDDK recognition site for human light chain enterokinase. The sample was incubated overnight with overexpressed human EK light chain purified from eCells. The signal at 25270.58 Da is the cleaved protein after the DDDDK site after 12 h of incubation at 37°C

resistance gene, MBP-DDDDK-EK under T7 promoter/terminator and a C112S mutation for stability, was transformed alongside a p15a vector with kanamycin resistance with PDI under a P_{BAD} promoter (P_{BAD} PDI/MBP-DDDDK-EK). 1 L of LB media was inoculated with a 10 mL starter culture of Xjb(DE3)* P_{BAD} PDI/EK-MBP. 3 mM of arabinose was added at the beginning of inoculation for the expression of PDI and endolysin and 1 g of eCells was made as described in [11]. 1 g of eCells was used with 20 mL creatine phosphate based CFPS expression at pH 7.5 with 1.7 mM GSH and 0.3 mM GSSG and incubated at 37°C overnight.

The soluble MBP-DDDDK-EK was expressed in its active form using eCell technology. 200 μg of protein was produced from the 20 mL eCell CFPS utilizing PDI and altered redox potential. To characterize whether the produced EK was active, 60 μl of 40 μM of red fluorescent protein (RFP) with a DDDDK motif preceding an N-terminal 6xHis tag (DDDDK-RFP) was incubated in cleavage buffer (20 mM Tris–HCl pH 8, 50 mM NaCl, 100 μM $CaCl_2$) with 6 μg of EK at 37°C O/N. The cleaved product of DDDDK-RFP was the major species in MS spectrum, although there was also non-specific cleavage of RFP at 25112 Da, which corresponds to cleavage at Ser10, with the 158 difference to corresponding to a product with Ser10 and Ala9 as seen in Fig. 5. The non-specific cleavage could be through EK, or another co-purified protease present within the EK sample. The original purified, uncleaved DDDDK-RFP (26813.08 Da) was not present in the MS, indicating a complete cleavage reaction.

A further demonstration where the eCell technology produced a difficult protein is human neuritin, a disulfide-bonded protein which has not been reported being

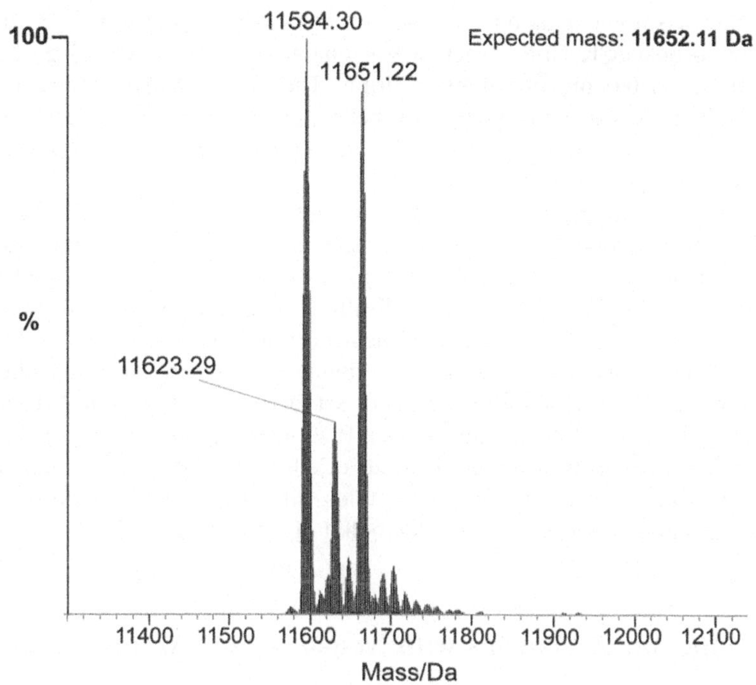

Fig. 6 High-resolution mass spectrum of Neuritin with a N-terminal 6xHIS tag and a TEV cleavage site. The signals show the neuritin with all the 6 cysteines forming disulfides. The expected mass is 11,652 Da for the disulfide-bonded form and 11,658 for reduced form. There are unidentified products at 11,594.30 and 11,623.29 Da

expressed successfully. Neuritin is a human neurotrophic factor which stimulates neuronal cell growth and is linked to inflammation and allergy. Neuritin is a disulfide-linked homodimer, consisting of two 9.72 kDa polypeptide monomer subunits. A pET vector with an ampicillin resistance gene, neuritin with a N-terminal 6xHIS tag and a TEV protease cleavage site under a T7 promoter/terminator was co-transformed into Xjb(DE3)* with a p15a vector with kanamycin resistance with PDI under a P_{BAD} promoter cells. A 5 mL CFPS was conducted with 300 mg of PDI overexpressed eCells as described in [11]. PDI is on an auxiliary P_{BAD} plasmid which expresses when induced with arabinose.

The soluble protein obtained from eCell expression was 72 μg from a 5 mL reaction and 300 mg of eCells. Intact mass spectrometry confirms that neuritin was produced with all three disulfides formed, however, other unidentified soluble products of neuritin at 11,594 and 11,623 Da were also detected in the MS (Fig. 6). The refolding of neuritin via protein disulfide isomerase (PDI) within the encapsulation during CFPS reaction was successful. In contrast, bacterial expression of neuritin produced protein in inclusion bodies, requiring refolding, and not guaranteeing the formation of the correct conformation of the protein. To further understand the structure of neuritin, the eCell expression conditions were scaled up to generate a sufficient amount of neuritin for NMR spectroscopy. A 100 mL creatine phosphate/CK CFPS reaction was conducted with about 3.5 g of eCells, which produced 0.6 mg of soluble neuritin. The 1D ^{1}H NMR spectrum of this neuritin sample, however, showed it was heterogeneous, with broad chemical shifts for amides, methyl groups, and sidechain protons, possibly pointing to fast internal Disulfide exchange in neuritin.

Typically, E. coli expression systems have issues producing Disulfide-rich proteins in a conformationally correct structure or in a soluble form, as the cytosol of *E. coli* is reducing, inhibiting the formation of disulfide bonds. The use of altered redox potential for the reshuffling of disulfides via PDI has been demonstrated previously with in vitro but not in complement to a CFPS reaction. The openness of the CFPS reaction allows for precise fine tuning of the redox potential in the CFPS reaction, enhancing PDI ability to form correct target protein structure. The use of eCells has the additional advantage that PDI does not have to be purified or specialized cell extracts need to be used. eCells is an inexpensive and easy to produce in vitro system that correctly folds difficult to produce proteins with multiple disulfide bonds, expanding the toolkit of synthetic biologists.

5 Production of Proteins with Non-standard Amino Acids

Using CFPS systems, site-specific protein phosphorylation has allowed the study of conformational shifts of proteins during phosphorylation events [29]. Threonine, serine, and tyrosine make up most phosphorylated residues, but it is difficult to specifically target individual residues for homogenous phosphorylation [30]. A reliable way to recreate phosphorylation at a specific residue is through site-specific non-canonical amino acid (ncAA) incorporation [31]. To incorporate a phosphorylated residue, an orthogonal amino-acyl tRNA synthetase (aaRS) and suppressor tRNA ($tRNA_{sup}$) pair is used to recode the amber (TAG) codon to introduce a phosphorylated ncAA into the polypeptide chain [32]. *E. coli* cell strains and $aaRS/tRNA_{sup}$ have been established for the incorporation of two phosphorylated ncAAs; phosphoserine (SeP) and phosphothreonine (pThr) [33, 34]. In naturally occurring phosphoproteins, these two residues are post-translationally installed by kinases, and make up over 98% of all phosphorylation occurrences [35]. To homogenously incorporate SeP and pThr translationally in *E.coli* cells, derivatives of BL21(DE3)* cell lines with knockouts for serB and serC were introduced. The knockout ΔserC prevents the production of phosphoserine (SeP) as the pThr aaRS/tRNAsup preferentially incorporates SeP rather than pThr, which is present in *E. coli* [34]. ΔserB prevents possible enzymatic dephosphorylation of SeP as the gene encoding phosphoserine phosphatase is inactivated [33]. To utilize these cell strains in site-specific protein phosphorylation in CFPS, producing a lysate based on these knockouts is essential.

Chemical modifications of proteins with toxic moieties are not possible to produce by in vivo expression. eCell technology is advantageous for complex ncAA incorporation, such as introducing site-specific phosphorylation and/or

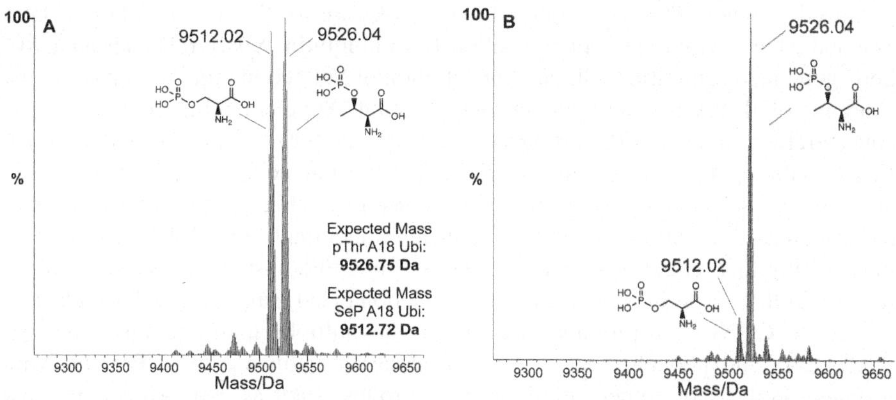

Fig. 7 (**a**) Mass spectrum of ubiquitin showing both O-phospho-L-threonine and O-phospho-L-serine incorporated at position 18, due to the high level of O-phospho-L-serine in using Xjb(DE3)* cells. (**b**) Mass spectrum of ubiquitin with only O-phospho-L-threonine incorporated at position 18 using functionalized ΔSerC eCells w/pLysS. High-resolution mass measurements were performed on an Orbitrap Elite mass spectrometer in positive mode at a resolution of 240,000

fluorination of alanine, which is toxic to cells [36]. Using eCells, we have incorporated several phosphorylated ncAAs that are difficult to express or have never been expressed in vitro. This expression has been done with different cell lines transformed with a pACYC-based plasmid carrying a T7 lysozyme gene derivative (pLysS) for eCell generation. This allows the functionalization of any gram-negative bacteria into encapsulated lysates (eCells), which then allow for their use in CFPS. In this case, we use BL21(DE3)* derivatives for the homogenous site-specific phosphorylation of L-phospho-O-threonine (pThr) in E18TAG ubiquitin. As a proof of principle we produced the protein streptococcal protein G 1 (GB1) with site-specific phosphorylation of a serine residue and simultaneously replaced all alanine residue with 3-fluoro-alanine, which is highly toxic to cells but does not affect gene transcription and ribosomal translation of proteins.

The first step was to establish whether pThr incorporation was possible in an in vitro system. Xjb(DE3)* cells were transformed with a pCloDF13 origin plasmid with a Ubiquitin gene under a T7 promoter which is amber interrupted at position 18 and with a 6xHis tag on the C-terminus (pCDF A18 Ubi) and the engineered pUC ThpRS tRNA(v2.0) pdux efSep previously reported. 1 g of eCells was produced from 1 L of LB media and then induced for CFPS expression using 20 mL CFPS buffer with 5 mM pThr supplemented. At the same time, the gene of a lytic enzyme on a biocompatible vector pLysS was transformed into BL21ΔserC with pCDF A18 Ubi alongside pUC ThpRS tRNA(v2.0) pdux efSep, to functionalize a non-autolysis strain of bacterium that has serC inactivated, preventing intracellular production of SeP. 1 g of BL21ΔserC eCells was produced from 1 L of LB media and then induced for CFPS expression using 20 mL CFPS buffer with 5 mM pThr supplemented.

The yield for XjB(DE3)* eCell expression was 0.95 mg, while BL21ΔserC was 0.72 mg. in Fig. 7a shows the mass spectrometry analysis of the XjB(DE3)* eCell produced samples. The two observed mass peaks are for the incorporation of both SeP and pThr in equal amount at position 18 of ubiquitin. XjB(DE3)* has an active SerC gene increasing intracellular concentration of SeP, competing with pThr in the charging of tRNA and subsequent incorporation. Supplementing pThr at 5 mM concentration in the CFPS produced protein with 50% of the desired product. Functionalizing BL21ΔserC into eCells avoid the intracellular production of SeP and allow for selective incorporation of pThr. As seen in Fig. 7b, the incorporation of pThr increased to approximately 90% using functionalized BL21ΔserC eCells, highlighting that reduction of intracellular SeP in CFPS systems increases incorporation and that bacterial cell lines can be functionalized using an auxiliary plasmid for eCell CFPS. To produce proteins with multiple modifications, such as fluorination and phosphorylation, it is necessary to utilize eCells. Proteins with residue-specific replacement of fluorinated groups, such as 3-fluoro-alanine, and phosphor-serine cannot be produced in vivo, due to cell toxicity and our inability to produce phosphoserine containing proteins in standard CFPS.

A vector with spectinomycin resistance, pCloDF13 origin with MASMTG-ENLFYQ-streptococcal protein G B1 gene (GB1) under a T7 promoter which is amber interrupted at position 24 and 28 was transformed into BL21ΔserB with the engineered phosphoserine orthogonal translation system (SepOTSλ) and PlysS. 0.6 g of BL21ΔserB eCells were used from this 3-plasmid system in a 15 mL CFPS with 4 mM SeP and 7 mM of DL-3-fluoro-alanine replacing L-alanine in the reaction mixture.

0.3 mg of protein was produced from 0.6 g of eCells of GB1 double incorporated SeP at A24 and A28. Figure 8 shows the mass spectrum of the produced protein. Both phosphoserine residues have been successfully introduced into the protein, but 3-fluoro-alanine replacement is only partially, with only one or two alanine residues, out of the possible three, replaced. These results suggest that alanine is either synthesized within eCells during the CFPS reaction or was partially retained in the eCells despite washing steps post-lysis. Using XjB(DE3)* cells for the introduction of SeP and 3-fluoro-alanine led to a mass spectrum that proved challenging to interpret. This complexity arose due to the susceptibility of the SeP to dephosphorylation. Consequently, a range of peaks representing both phosphorylated and dephosphorylated forms emerged, obscuring the analysis. Moreover, the replacement of alanine residues with 3-fluoro-alanine was incomplete, further complicating the interpretation of the mass spectrum.

The functionalization of *E. coli* derivatives into eCells using a lytic auxiliary plasmid has allowed for homogenous phosphorylation of different protein targets. In combination, the residue-specific fluorination of alanine has been applied to protein targets. This allows chemically interesting derivatives of amino acids to be incorporated into proteins for NMR spectroscopy, as ^{19}F is an attractive NMR Active spin probe? which can be detected readily with high sensitivity. The incorporation of native alanine alongside 3-fluoro-alanine highlights the issue of natural amino acids being present in the eCells prior to expression. Extensive dialysis steps have been applied, but important macromolecules related to transcription and translation can leak over time from the permeable encapsulation, ultimately reducing protein yield.

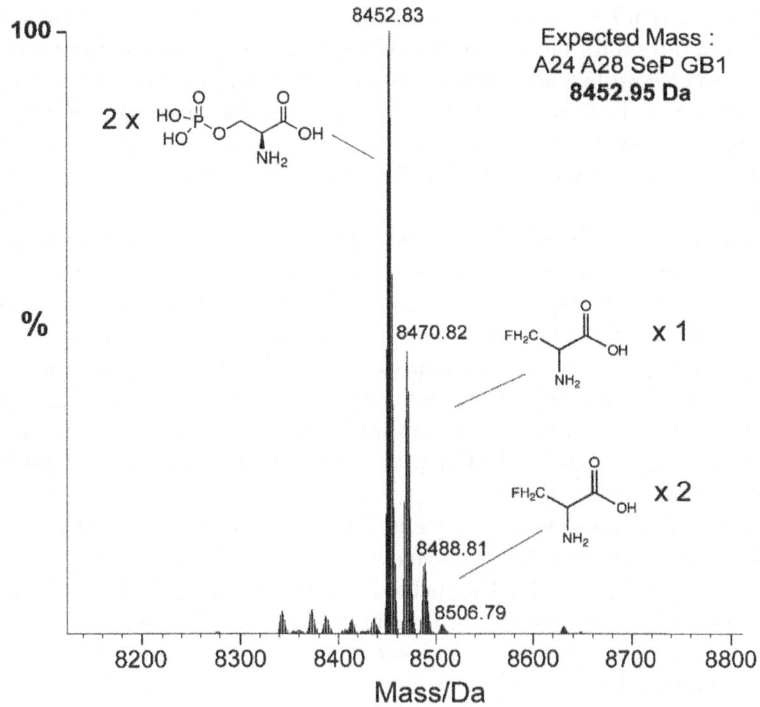

Fig. 8 High-resolution mass spectrum of GB1 containing two phosphor-serine residues and alanine being replaced by 3-fluoro-alanine. The signals show the measure masses of the molecules and the expected (calculated) masses for GB1 with two phosphor-serine (SeP) and varying number of 3-fluoro-alanine replacing native alanine. This expression was done using functionalized ΔSerB eCells w/pLysS, 4 mM SeP and 7 mM 3-fluoro-alanine. The expression yield was or 0.4 mg from a 15 mL eCFPS reaction

An auxiliary plasmid (pLysS) enables the expansion of eCells to any gram-negative bacterial cell line, such as XjB(DE3)*, facilitating the production of capsules. We demonstrate this by functionalizing two *E. coli* cell lines, BL21ΔserC and BL21ΔserB for incorporation of SeP and pThr.

6 Summary

In this study, an in vitro expression system was employed using eCells to detect heavy metal ions, specifically mercury. The eCell system was shown to successfully detect mercury ions at higher sensitivity compared to in vivo expression. The semipermeable nature of the eCells capsules allows for passive diffusion of the metal ions, improving stringency. The eCells system is efficient, inexpensive, and can be stored for extended periods of time, making it a portable and accessible biosensing technology without risk to the environment. The eCell technology is a promising approach for bioremediation of toxicants in extreme environments due to its ability to encapsulate and protect enzymes, providing greater stability and shelf life compared to isolated or purified enzymes. Furthermore, the eCells exhibit robustness to environmental changes such as high salinity and pH, and their lyophilization feature enhances their portability, making them a viable option for on-site bioremediation.

Additionally, eCell technology can be used for expressing disulfide-bonded proteins in a soluble, active form. This is accomplished by controlling the expression environment and overexpressing protein disulfide isomerase (PDI) to mimic the redox potential of the endoplasmic reticulum. The semipermeable capsule of the eCell system allows for the retention of both the protein and PDI, providing a controlled environment for conformationally correct protein folding. The efficacy of this technology was demonstrated using human light chain enterokinase, and a human neurotrophic factor called neuritin, which produced significant amounts of protein for NMR spectroscopy.

Moreover, this study presents a method for simultaneous incorporation of SeP, pThr, and 3-fluoro-alanine into proteins using eCell CFPS with *E. coli* derivatives and a lytic auxiliary plasmid. The method enables site-specific phosphorylation of different protein targets and the incorporation of chemically interesting derivatives of amino acids. The approach is quicker and less costly than other methods for producing capsule lysate.

In conclusion, eCell technology is a versatile and promising approach for various biotechnological applications. It has shown to be efficient, cost-effective, and environmentally friendly, with potential use in biosensing and bioremediation, as well as in the production of challenging-to-express proteins.

References

1. Carlson ED, Gan R, Hodgman CE, Jewett MC (2012) Cell-free protein synthesis: applications come of age. Biotechnol Adv 30(5):1185–1194. https://doi.org/10.1016/J.BIOTECHADV.2011.09.016
2. Shrestha P, Holland TM, Bundy BC (2012) Streamlined extract preparation for Escherichia coli-based cell-free protein synthesis by sonication or bead vortex mixing. Biotechniques 53(3): 163–174. https://doi.org/10.2144/0000113924
3. Apponyi MA, Ozawa K, Dixon NE, Otting G (2008) Cell-free protein synthesis for analysis by NMR spectroscopy. Methods Mol Biol 426:257–268. https://doi.org/10.1007/978-1-60327-058-8_16
4. Kim TW, Keum JW, Oh IS, Choi CY, Kim HC, Kim DM (2007) An economical and highly productive cell-free protein synthesis system utilizing fructose-1,6-bisphosphate as an energy source. J Biotechnol 130(4):389–393. https://doi.org/10.1016/J.JBIOTEC.2007.05.002
5. Kim TW, Kim HC, Oh IS, Kim DM (2008) A highly efficient and economical cell-free protein synthesis system using the S12 extract of Escherichia coli. Biotechnol Bioprocess Eng 13(4): 464–469. https://doi.org/10.1007/S12257-008-0139-8

6. Jewett MC, Swartz JR (2004) Mimicking the Escherichia coli cytoplasmic environment activates long-lived and efficient cell-free protein synthesis. Biotechnol Bioeng 86(1):19–26. https://doi.org/10.1002/BIT.20026
7. Perez JG, Stark JC, Jewett MC (2016) Cell-Free synthetic biology: engineering beyond the cell. Cold Spring Harb Perspect Biol 8(12). https://doi.org/10.1101/CSHPERSPECT.A023853
8. Liu WQ, Wu C, Jewett MC, Li J (2020) Cell-free protein synthesis enables one-pot cascade biotransformation in an aqueous-organic biphasic system. Biotechnol Bioeng 117(12): 4001–4008. https://doi.org/10.1002/BIT.27541
9. Katzen F, Chang G, Kudlicki W (2005) The past, present and future of cell-free protein synthesis. Trends Biotechnol 23(3):150–156. https://doi.org/10.1016/J.TIBTECH.2005.01.003
10. Hillberg AL, Tabrizian M (2006) Biorecognition through layer-by-layer polyelectrolyte assembly: in-situ hybridization on living cells. Biomacromolecules 7(10):2742–2750. https://doi.org/10.1021/bm060266j
11. Van Raad D, Huber T (2021) In vitro protein synthesis in semipermeable artificial cells. ACS Synth Biol 10(5):1237–1244. https://doi.org/10.1021/ACSSYNBIO.1C00044/SUPPL_FILE/SB1C00044_SI_001.PDF
12. Hunt JP, Galiardi J, Free TJ, Yang SO, Poole D, Zhao EL et al (2022) Mechanistic discoveries and simulation-guided assay optimization of portable hormone biosensors with cell-free protein synthesis. Biotechnol J 17(2):e2100152. https://doi.org/10.1002/biot.202100152
13. Lu Y (2022) Textile-embedded cell-free biosensors. Nat Biomed Eng 6(3):225–226. https://doi.org/10.1038/s41551-022-00869-3
14. Salehi AS, Shakalli Tang MJ, Smith MT, Hunt JM, Law RA, Wood DW et al (2017) Cell-free protein synthesis approach to biosensing hTRβ-specific endocrine disruptors. Anal Chem 89(6): 3395–3401. https://doi.org/10.1021/acs.analchem.6b04034
15. Soltani M, Davis BR, Ford H, Nelson JAD, Bundy BC (2018) Reengineering cell-free protein synthesis as a biosensor: biosensing with transcription, translation, and protein-folding. Biochem Eng J 138:165–171. https://doi.org/10.1016/j.bej.2018.06.014
16. Free TJ, Tucker RW, Simonson KM, Smith SA, Lindgren CM, Pitt WG et al (2023) Engineering at-home dilution and filtration methods to enable paper-based colorimetric biosensing in human blood with cell-free protein synthesis. Biosensors 13(1):104
17. Wang X, Cheng M, Yang Q, Wei H, Xia A, Wang L et al (2019) A living plant cell-based biosensor for real-time monitoring invisible damage of plant cells under heavy metal stress. Sci Total Environ 697:134097. https://doi.org/10.1016/j.scitotenv.2019.134097
18. Jung JK, Alam KK, Verosloff MS, Capdevila DA, Desmau M, Clauer PR et al (2020) Cell-free biosensors for rapid detection of water contaminants. Nat Biotechnol 38(12):1451–1459. https://doi.org/10.1038/s41587-020-0571-7
19. Gupta N, Renugopalakrishnan V, Liepmann D, Paulmurugan R, Malhotra BD (2019) Cell-based biosensors: recent trends, challenges and future perspectives. Biosens Bioelectron 141: 111435. https://doi.org/10.1016/j.bios.2019.111435
20. Wan X, Volpetti F, Petrova E, French C, Maerkl SJ, Wang B (2019) Cascaded amplifying circuits enable ultrasensitive cellular sensors for toxic metals. Nat Chem Biol 15(5):540–548. https://doi.org/10.1038/s41589-019-0244-3
21. Davison J (2005) Risk mitigation of genetically modified bacteria and plants designed for bioremediation. J Ind Microbiol Biotechnol 32(11–12):639–650. https://doi.org/10.1007/s10295-005-0242-1
22. Karig DK (2017) Cell-free synthetic biology for environmental sensing and remediation. Curr Opin Biotechnol 45:69–75. https://doi.org/10.1016/j.copbio.2017.01.010
23. Sales da Silva IG, Gomes de Almeida FC, Padilha da Rocha e Silva NM, Casazza AA, Converti A, Asfora Sarubbo L (2020) Soil bioremediation: overview of technologies and trends. Energies 13(18):4664
24. Padhan D, Rout PP, Kundu R, Adhikary S, Padhi PP (2021) Bioremediation of heavy metals and other toxic substances by microorganisms. Soil Bioremediation:285–329

25. Campbell EC, Grant J, Wang Y, Sandhu M, Williams RJ, Nisbet DR et al (2018) Hydrogel-immobilized supercharged proteins. Advanced. Biosystems 2(7):1700240. https://doi.org/10.1002/adbi.201700240
26. Singh A, Upadhyay V, Upadhyay AK, Singh SM, Panda AK (2015) Protein recovery from inclusion bodies of Escherichia coli using mild solubilization process. Microb Cell Fact 14(1): 41. https://doi.org/10.1186/s12934-015-0222-8
27. Moghadam M, Ganji A, Varasteh A, Falak R, Sankian M (2015) Refolding process of cysteine-rich proteins: Chitinase as a model. Rep Biochem Mol Biol 4(1):19–24
28. Martens E, Alloza I, Scott CJ, Billiau A, Vandenbroeck K (2000) Protein disulfide isomerase-mediated cell-free assembly of recombinant interleukin-12 p40 homodimers. Eur J Biochem 267(22):6679–6683. https://doi.org/10.1046/j.1432-1327.2000.01765.x
29. Nishi H, Shaytan A, Panchenko AR (2014) Physicochemical mechanisms of protein regulation by phosphorylation. Front Genet:5. https://doi.org/10.3389/fgene.2014.00270
30. Silvestroni A, Jewell KA, Lin WJ, Connelly JE, Ivancic MM, Tao WA et al (2009) Identification of serine/threonine kinase substrates in the human pathogen group B streptococcus. J Proteome Res 8(5):2563–2574. https://doi.org/10.1021/pr900069n
31. Qin X, Liu T (2022) Recent advances in genetic code expansion techniques for protein phosphorylation studies. J Mol Biol 434(8):167406. https://doi.org/10.1016/j.jmb.2021.167406
32. Chen J, Tsai YH (2022) Applications of genetic code expansion in studying protein post-translational modification. J Mol Biol 434(8):167424. https://doi.org/10.1016/j.jmb.2021.167424
33. Rogerson DT, Sachdeva A, Wang K, Haq T, Kazlauskaite A, Hancock SM et al (2015) Efficient genetic encoding of phosphoserine and its nonhydrolyzable analog. Nat Chem Biol 11(7): 496–503. https://doi.org/10.1038/NCHEMBIO.1823
34. Zhang MS, Brunner SF, Huguenin-Dezot N, Liang AD, Schmied WH, Rogerson DT et al (2017) Biosynthesis and genetic encoding of phosphothreonine through parallel selection and deep sequencing. Nat Methods 14(7):729–736. https://doi.org/10.1038/NMETH.4302
35. Honkanen ER, Golden T (2002) Regulators of serine/threonine protein phosphatases at the Dawn of a clinical era? Curr Med Chem 9(22):2055–2075. https://doi.org/10.2174/0929867023368836
36. Kollonitsch J, Barash L (1976) Organofluorine synthesis via photofluorination: 3-fluoro-D-alanine and 2-deuterio analogue, antibacterials related to the bacterial cell wall. J Am Chem Soc 98:5591–5593

The manufacturer's authorised representative in the EU is Springer Nature Customer Service Centre GmbH, Europaplatz 3, 69115 Heidelberg, Germany. If you have any concerns regarding our products, please contact ProductSafety@springernature.com

Printed and bound by CPI Group (UK) Ltd, Croydon, CR0 4YY

28/09/2026

02224074-0017